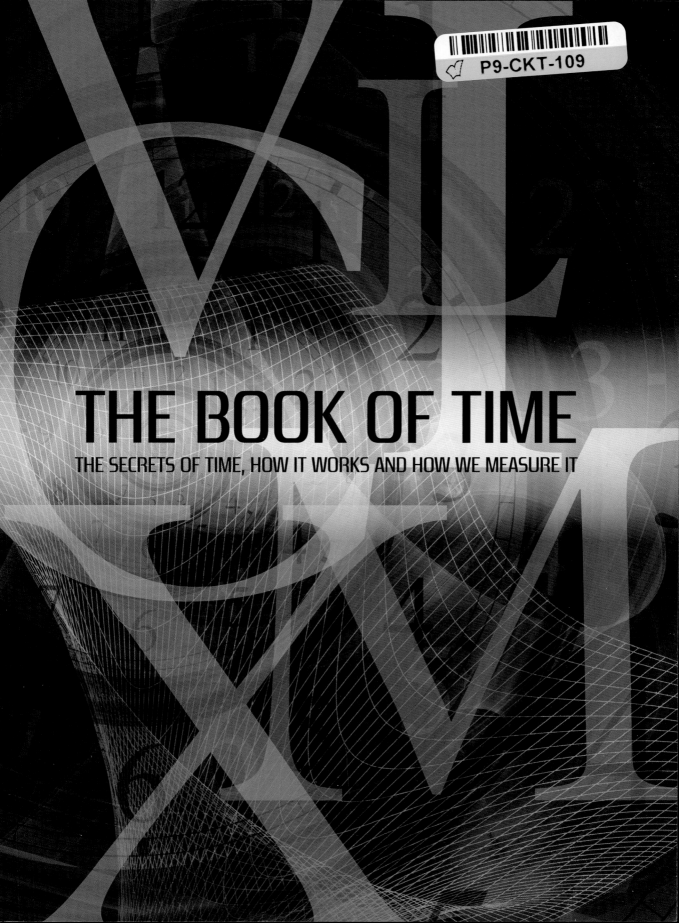

THE BOOK OF TIME

THE SECRETS OF TIME, HOW IT WORKS AND HOW WE MEASURE IT

A FIREFLY BOOK

Published by Firefly Books Ltd. 2011

First printing

Publisher Cataloging-in-Publication Data (U.S.)
Hart-Davis, Adam.
The book of time : the secret of time, how it works and how we measure it / Adam Hart-Davis.
[256] p. : ill. (some col.), col. photos. ;   cm.
Includes index.
Summary: Guide that explores every aspect of time and how we measure it, including relationships with the natural world, our bodies, religion, philosophy, social and political history, science, psychology, physics, astronomy, commerce.
ISBN-13: 978-1-55407-905-6 (pbk.)
1. Time. 2. Time measurements. I. Title.
529.7 dc22   QB213.H367   2011

Library and Archives Canada Cataloguing in Publication
Hart-Davis, Adam
The book of time : the secret of time, how it works and how we measure it / Adam Hart-Davis.
Includes index.
ISBN 978-1-55407-905-6
1. Time. I. Title.
QB209.H37 2011   529   C2011-904067-0

Published in the United States by
Firefly Books (U.S.) Inc.
P.O. Box 1338, Ellicott Station
Buffalo, New York 14205

Published in Canada by
Firefly Books Ltd.
66 Leek Crescent
Richmond Hill, Ontario L4B 1H1

Printed in China

Credits
Back cover (left to right): Hulton-Deutsch Collection/Corbis; Pseudolongino/Dreamstime.com; courtesy Wikipedia Commons
Front cover: Digital Art/Corbis (world with arrows); Image Source/Corbis (clock)
Front cover design: Martin Gould

ADAM HART-DAVIS

THE BOOK OF TIME

THE SECRETS OF TIME, HOW IT WORKS AND HOW WE MEASURE IT

FIREFLY BOOKS

CONTENTS

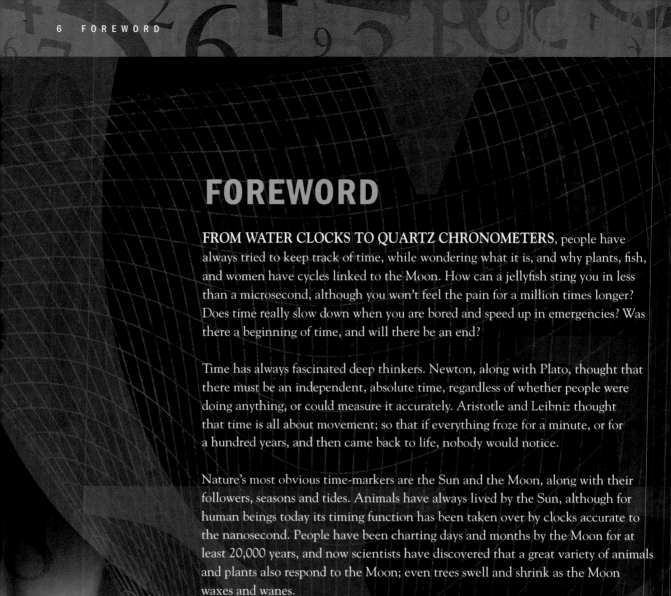

FOREWORD

FROM WATER CLOCKS TO QUARTZ CHRONOMETERS, people have always tried to keep track of time, while wondering what it is, and why plants, fish, and women have cycles linked to the Moon. How can a jellyfish sting you in less than a microsecond, although you won't feel the pain for a million times longer? Does time really slow down when you are bored and speed up in emergencies? Was there a beginning of time, and will there be an end?

Time has always fascinated deep thinkers. Newton, along with Plato, thought that there must be an independent, absolute time, regardless of whether people were doing anything, or could measure it accurately. Aristotle and Leibniz thought that time is all about movement; so that if everything froze for a minute, or for a hundred years, and then came back to life, nobody would notice.

Nature's most obvious time-markers are the Sun and the Moon, along with their followers, seasons and tides. Animals have always lived by the Sun, although for human beings today its timing function has been taken over by clocks accurate to the nanosecond. People have been charting days and months by the Moon for at least 20,000 years, and now scientists have discovered that a great variety of animals and plants also respond to the Moon; even trees swell and shrink as the Moon waxes and wanes.

Organizing the time and the date has provided ways to control people over the centuries. Nilometers, calendar wheels, and Pope Gregory's advisers have all contributed to the way days and dates and years have been marked and recorded, and now our year of 365.2419 days is only 27 seconds out of sync with the Sun.

How can time be measured with such precision? People have devised and improved clocks for thousands of years, and perfected longcase timekeepers in the 17th and 18th centuries. The atomic and electronic beasts of today may look like boring boxes, but they are far more measured and precise in their ticking than any of the elegant mechanical movements of yesteryear, or even the Earth itself.

And finally, how does science view time? Einstein told us that time is relative, and depends on your point of view. Many scientists tell us that time travel into the future is possible – indeed we all do it, at one second per second – but one Russian has travelled much further. Into the past? Maybe, but that is speculation, and runs into the grandmother paradox.

How complex it all is, but in this book I hope I have made some bits easy to understand, and I hope you enjoy finding out about time as much as I have.

Adam Hart-Davis

CHAPTER 1

The short answer is "time is the stuff that is measured by clocks", but that is only a small part of the answer. For one thing, only humans have clocks (as far as we know), and surely other animals have some idea of time, even if only night and day. So unless time is a purely human construct, it must have some existence outside clocks. Or must it? Philosophers have argued for centuries about the nature of time, and some have asserted that it does not exist at all. Meanwhile our perception of time seems to depend as much on the circumstances as on the passing seconds.

The Greek philosophers had a variety of firm views about time. In about 500 BC Heraclitus wrote, obscurely, "On those stepping into rivers the same, other and other waters flow," which Plato interpreted as "Everything changes and nothing remains still," or "You can never step twice into the same river."

THE ANCIENT GREEKS

ABOVE: A bowl or jar with a hole in the bottom is the simplest form of water clock. The water takes a fixed time to run out. Early Greek designs were based on Egyptian clocks, like this one. Later, top-up systems and gears were added.

PLATO HIMSELF, BORN IN ABOUT 428 BC and founder of the Academy in Athens, the western world's first "university", asserted that time had been created along with the universe, and has an independent existence. If everything were to freeze and remain unchanged and unmoving for a second, a minute, or a hundred years, and then carry on as before, no one would notice – but time would have moved on. Plato thought of time as an empty container that could be filled with objects and events, and through which they can move.

　　Plato's pupil Aristotle was to become the most famous philosopher of all time. He was born in 384 BC, studied at Plato's Academy for 20 years, taught the son of King Philip II of Macedon, who became Alexander the Great, and wrote books on virtually every subject then known. He may well have been the last person on Earth to have known everything that there was to be known at the time. Aristotle returned to Athens in 335 BC and founded his own school, the Lyceum, where he taught for 12 years.

Aristotle disagreed with Plato, and asserted that time is all about change and movement – obvious in stars and the water-clocks they used – and therefore that if everything were to freeze then time would stop too. Time, he said, does not exist independently of events, which means that time without change is meaningless. This is reductionism: all discussion about time can be reduced to the events that are happening.

Is time *ABSOLUTE, ETERNAL, and EXTERNAL...* or does it depend on us?

Zeno's paradoxes

In about 450 BC, three or four generations before Aristotle, Parmenides had asserted that time and motion were not real, but merely illusions. His pupil Zeno of Elea introduced the idea of proof by *reductio ad absurdum*, which he used in some of his famous paradoxes. Among the best known of these are "Achilles and the Tortoise", and the "Arrow paradox".

BELOW: Plato (left) and Aristotle (right) at the School of Athens in a famous painting by Raphael (1510). Plato is pointing upwards, towards the eternal; Aristotle has his sights on the world in front of him.

Zeno imagined that Achilles, a hero of the Trojan War and a notably fast runner, has challenged a tortoise to a running race. Achilles gives the tortoise start of a few metres; so the tortoise starts from point T1. They start at the same moment, and normal experience suggests that Achilles should rapidly pass the tortoise and win the race. According to Zeno, however, Achilles takes a certain amount of time (say 1 second) to cover the first few metres and reach T1, which is where the tortoise started from. By that time the tortoise has crawled a little way ahead, and has reached point T2. Achilles will take a little longer to reach T2, by which time the tortoise will have reached T3, and so on. No matter how fast he runs, Achilles will never actually catch the tortoise. According to the epic Greek poem *The Iliad*, Achilles was the only mortal to be consumed by rage; so presumably he would have been exceedingly annoyed at being unable to catch the tortoise.

The Arrow Paradox is perhaps more confusing. Zeno said that an arrow in flight must at any one instant of time occupy a particular position in space. If it is moving it must either move to where it is, or it must move to where it is not. But it can't move to where it is not in this instant, and it can't move to where it is, because it's there already. Therefore the arrow must be stationary, and so it cannot be moving. This certainly worried Aristotle, who wrote in his treatise on physics, "If everything when it occupies an equal space is at rest, and if that which is in locomotion is always occupying such a space at any moment, the flying arrow is therefore motionless."

If, during a fragment of time, the arrow moved, then that fragment of time would be divisible. Aristotle came to the conclusion that time is not made of a sequence of indivisible "Now" moments, but rather is itself movement, and flows, like a river, or an arrow across the sky.

ABOVE: The Greek philosopher Zeno was interested in the relationship between time and space. One of his famous paradoxes "proves" that a tortoise, given a head start, cannot be overtaken, even by a champion runner.

IS TIME ABSOLUTE?

This is perhaps the deepest question about time: is it absolute, external, and independent of us, the world, and everything, or is it only about things happening? Was Plato right to think that if everything froze for 100 years no one would notice? A parallel question, which may help us frame questions about time, concerns mathematics. Is mathematics absolute, and was it inherent in the universe, waiting to be discovered by mathematicians, or did the mathematicians invent it?

Now?

Past, present, and future are the constituents of time. The past is history; the present is now; the future is yet to come. "Now" is often called "the moving present" – it connects past and future and sweeps through our lives and the universe. But what is this Now? Is it an infinitely thin slice of time – the smallest imaginable unit? Does time consist of an infinite procession of these Nows?

Aristotle thought not. At any given moment Now, the previous Now must have disappeared. It cannot have disappeared in its own lifetime,

for then it was Now; but it cannot have disappeared in a subsequent Now, since the Nows must be sequential and cannot coexist. Therefore it seems that there is no succession of Nows. Furthermore, some segments of time have already passed, and others are to come, but none of them is Now, for Now cannot be a segment; it is only the division between past and future. It follows, some say, that time itself does not exist, since neither past nor present exists, and if there is no Now, then there is no time.

Augustine of Hippo (more of him later) discussed the knife-edge of Now: "If any portion of time be conceived which cannot now be divided into even the minutest particles of moments, this only is that which may be called present; which, however, flies so rapidly from future to past, that it cannot be extended by any delay. For if it be extended, it is divided into the past and future; but the present hath no space."

ABOVE: Zeno's arrow paradox seems to "prove" that a moving object is really stationary. He imagined time as a sequence of instants, in none of which is the arrow actually moving.

Aurelius Augustinus, born in AD 354 in what is now Algeria, had a profound crisis at the age of 32 and converted to Christianity. In AD 395 he was made a bishop of Hippo Regius (now Annaba) in northeast Algeria, and became known as Augustine of Hippo.

Augustine of Hippo

IN HIS CONFESSIONS (AD 397–98), Augustine ponders the nature of time at considerable length in the form of a prayer or a series of questions to God. Here are a few excerpts:

"What is time? Who can easily and briefly explain it? If no one ask of me, I know; if I wish to explain to him who asks, I know not. Yet I say with confidence, that I know that if nothing passed away, there would not be past time; and if nothing were coming, there would not be future time; and if nothing were, there would not be present time. Those two times, therefore, past and future, how are they, when even the past now is not; and the future is not as yet?

"All time past is forced on by the future, and that all the future followeth from the past, and that all, both past and future, is created and issues from that which is always present."

Augustine also discusses the apparent paradoxes of past and future:

"If there are times past and future, I desire to know where they are. But if as yet I do not succeed, I still know, wherever they are, that they are not there as future or past, but as present. For if there also they be future, they are not as yet there; if even there they be past, they are no longer there. Wheresoever, therefore, they are, whatsoever they are, they are only so as present. Although past things are related as true, they are drawn out from the memory – not the things themselves, which have passed, but the words conceived from the images of the things which they have formed in the mind as footprints in their passage through the senses. My childhood, indeed, which no longer is, is in time past, which now is not; but when I call to mind its image, and speak of it, I behold it in the present, because it is as yet in my memory.

"Whether there be a like cause of foretelling future things, that of things which as yet are not the images may be perceived as already existing, I confess, my God, I know not. This certainly I know, that we generally think before on our future actions, and that this premeditation is present; but that the action whereon we premeditate is not yet, because

it is future; which when we shall have entered upon, and have begun to do that which we were premeditating, then shall that action be, because then it is not future, but present.

"I behold daybreak; I foretell that the sun is about to rise. That which I behold is present; what I foretell is future — not that the sun is future, which already is; but his rising, which is not yet. Yet even its rising I could not predict unless I had an image of it in my mind, as now I have while I speak. But that dawn which I see in the sky is not the rising of the sun, although it may go before it, nor that imagination in my mind; which two are seen as present, that the other which is future may be foretold. Future things, therefore, are not as yet; and if they are not as yet, they are not. And if they are not, they cannot be seen at all; but they can be foretold from things present which now are, and are seen.

"But what now is manifest and clear is, that neither are there future nor past things. Nor is it fitly said, "There are three times, past, present, and future;" but perchance it might be fitly said, "There are three times; a present of things past, a present of things present, and a present of things future." For these three do somehow exist in the soul, and otherwise I see them not: present of things past, memory; present of things present, sight; present of things future, expectation.

"I have heard from a learned man that the motions of the sun, moon, and stars constituted time, and I assented not. For why should not rather the motions of all bodies be time? What if the lights of heaven should cease, and a potter's wheel run round, would there be no time by which we might measure those revolutions."

RIGHT: Augustine, in a painting by Botticelli (c.1480), takes a break from his ecumenical duties to ponder time.

One of the first to think of time scientifically was Isaac Newton, who realized he needed it as a variable for his equations. Born on Christmas Day 1642, Newton studied at Cambridge, where he was influenced by the work of William Charleton, who had written, in 1654, that time was a real entity, that it "flows on eternally in the same calm and equal tenor", and that it is distinct from any measure of it.

NEWTON, LEIBNIZ, AND KANT

BELOW: Sir Isaac Newton realized that motion was relative – a person walks at a speed relative to the ground, and the Earth moves in orbit relative to the Sun. But he thought there was an absolute time and an absolute space against which all motion could ultimately be measured.

NEWTON DID HIS FINEST WORK in the plague years of 1665 and 1666, when Cambridge University was closed and he stayed at home in Lincolnshire, in the east of England. Pondering alone, he solved several mathematical problems, investigated the colours of the spectrum as produced by a prism and in rainbows, and had the first inspiration about gravity. His greatest work, and perhaps the most important science book of all time, was the 1687 *Principia*, or *Philosophiae Naturalis Principia Mathematica*, to give it the full title. He wrote it in Latin; it was translated by Andrew Motte in 1729 and revised by Florian Cajori in 1934.

In his *Principia* Newton laid down the laws of motion and the basis of physics for the next 300 years. The concept of time was fundamental to his arguments, and in the *Scholium* (introduction) he went out of his way to define what he meant by time: "Absolute, true, and mathematical time, of itself, and from its own nature, flows equably without relation to anything external, and by another name is called duration: relative, apparent, and common time, is some sensible and external (whether accurate or unequable) measure of duration by the means of motion, which is commonly used instead of true time; such as an hour, a day, a month, a year."

RIGHT: Newton's theory of gravity provided one rule that explained not only the movements of the moon and planets but also the behaviour of falling objects on Earth.

THE COLOURS OF THE RAINBOW

In 1666 Newton bought a glass prism at a country fair and was fascinated by the beautiful colours it produced from a beam of sunlight. Wondering how and why these colours appeared, he did a series of experiments, and discovered that sunlight is actually made up of a mixture of all the colours of the rainbow. All a prism does is separate them again. In a letter to the Royal Society he explained all this, and also described the design of a reflecting telescope that would allow astronomers to look at stars and planets without seeing coloured fringes.

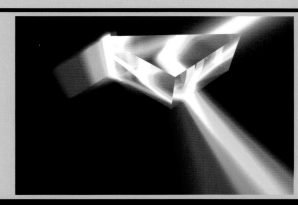

In other words, Newton distinguishes carefully between absolute time and common time. Absolute or mathematical time ticks on continuously, without any outside influence, and regardless of events or measurements. In this realist view Newton considers time to be a fundamental part of the universe. As he later says, "Absolute time, in astronomy, is distinguished from the relative, by the equation or correction of the apparent time. For the natural days are truly unequal, though they are commonly considered as equal, and used for a measure of time; astronomers correct this inequality that they may measure the celestial motions by a more accurate time. It may be that there is no such thing as an equable motion, whereby time may be accurately measured. All motions may be accelerated and retarded, but the flowing of absolute time is not liable to any change."

Gottfried Leibniz

A German contemporary of Newton's, Gottfried Wilhelm Leibniz, who called himself a Baron, invented calculus independently of Newton, though Newton accused him of plagiarism. Leibniz argued that God is rational, and needed a specific reason for everything. Therefore He would have needed a specific reason to create the universe at one particular time. Newton said every moment

Despite discovering the laws of motion and **GRAVITY**

Newton had a secret passion for alchemy and was extremely religious

TIME
is what happens when nothing else does
Richard Feynman

was identical to every other moment. So God would have had no reason to create the universe; therefore, Leibniz reasoned, Newton must be wrong.

Leibniz also suggested that everything existed as a sort of thread, or perhaps like a caterpillar, with each segment joined in time to the segments before and after, so that the entire existence of the thing is an actual object in space-time. He also believed that without events there would be no time; in other words he agreed with Aristotle that time is all about events and change, whereas Newton was on the side of Plato, and thought of time as ticking on quite independently.

Between 1715 and 1716, Leibniz and the English clergyman Samuel Clarke had a long correspondence about the nature of time, not to mention space, gravity, and miracles; and since Clarke was a close friend of Newton the ideas may well have come from him. Clarke (Newton) argued that space and time are real and fixed. Spin a bucket of water, he said, and the water will climb up the sides of the bucket. This could not happen unless the bucket were spinning in a fixed frame of reference; if the rest of the universe were spinning with the bucket, the water would not climb. Leibniz disagreed sharply. He said an empty space would be a substance with no properties.

Immanuel Kant

The German philosopher Immanuel Kant, in his *Critique of Pure Reason* (1781), says that time is a basic intuition that allows us to understand the experience of our senses. Time is not a substance, but part of a systematic framework that we need to understand anything.

A PERENNIAL QUESTION

At the end of the 17th century arose once again that profound question: Is time independent and absolute, or does it depend on us? Plato & Newton v Aristotle & Leibniz. Whose side are you on?

LEFT: Immanuel Kant lived his life in a timely fashion. His daily walk was such a fixed routine that townsfolk would set their clocks when he appeared.

OPPOSITE: A bucket spinning when it is raised from a well may spill as the water climbs up the sides. Newton would say the water climbs because it is spinning in a fixed frame of reference.

ABOVE: The behaviour of gravity at the subatomic, quantum level, according to theoretical physicist Lee Smolin, seems to confirm that time is real, and not some abstract construct.

MIDDLE: Our memories are not stopwatches. Without a scientific instrument we can only make subjective estimates of time duration.

part of last month. But no time can be both past and future; so the A series is self-contradictory, and therefore time is unreal. This sounds to me like a semantic argument, but I am not a philosopher.

In a 2010 magazine, American theoretical physicist Lee Smolin argued that time is real. Time is not, as many diverse thinkers have argued, from ancient Greeks through to contemporary quantum cosmologists, an illusion, an "artefact of our psychology". Nor does it lie in some timeless realm of truth, unconnected to the flow of the events we perceive. Smolin believes that current advances in the study of quantum gravity confirm, if anything, that our four-dimensional space-time only seems to make sense if we assume that time is real.

THE TOPOLOGY OF TIME

The timeline of an object is the path of its passage through time, but is this timeline just a straight line, or is it branched, so that different futures diverge? Could it be looped, and come back to the same time at some point in the future (which would be a form of time travel)? Or is it perhaps intermittent, so that instants or even longer periods are missing? Aristotle argued that time cannot have had a beginning, since the first moment must have come between some bit of past and the future, and therefore could not be the first moment.

In his words, "I would hold that, contrary to the ancient metaphysical tradition, time is not only real, it is likely that it is the only aspect of reality we experience directly that is fundamental and not emergent from anything else."

The psychology of time

Perceiving time is curious. We perceive the world around us with our senses, but we can't see time, nor touch it, nor taste, nor smell it. Even without using any of these senses, we can sense time going past by our train of thoughts. We can see the diver entering the water, but the only way we know she has left the diving board is by recalling memory; we cannot directly perceive the interval in time. This seems to obscure the entire subject. How can we hope to explain what time is if we cannot perceive it?

In 1978 German psychologist Ernst Pöppel listed various "elementary time experiences", including duration, non-simultaneity, order, past and present, and change. Duration seems to be simple – how long does a firework, say, take to finish fizzing? But without scientific apparatus (a stop-watch) this depends on memory, and cannot be measured before the event is completed.

ABOVE: There are a number of distinct "time experiences". Estimating some fairly short durations seems to be easy. For example, we can tell rough how long a firework takes burn out.

0,2 s

3
6
9
2
5
10
15
20

If your **BRAIN** *registers several events in a short time, it may make up a story to account for them, but this may not correspond with what happened*

To lead a normal life we need a notion of pastness. If consciousness were a string of un con nec t ed sensations, we would have no knowledge except of the instant

BELOW: Lingering memories
and fresh, incoming
experiences make up
the human stream of
consciousness, with each
sequence falling away
into the past.

Events in the near past may linger in a sort of memory trace, which would allow you to estimate both how long they lasted and how long ago they occurred – "that meal last night was delicious, although the service was slow". As time passes the memories fade, and in principle you might be able to work out the details from the strength of the trace. However, some memories fade more quickly than others, and you may instead have to rely on other events against which to map the one in question. My cat is about 15 years old, but I cannot remember his age directly. I can remember how he came to our family, but I cannot work out the duration of his life without external reference. We actually acquired him as a young kitten to cheer up my partner, who was ill at the time, and we know when she was ill; so indirectly we know the age of the cat.

Order is not always simple; when different events – like a flash and a bang – happen in rapid succession, it may be obvious that they were not simultaneous, but not which happened first. If your brain registers several events happening in a short space of time it may make up a plausible story to account for them – but the story may not correspond with what really happened.

Look at the second hand on a watch or clock, and you will see it moving. What you actually perceive, at most instants, is a stationary hand, but in your memory you hold an image of the hand in its previous position, and perhaps of the blur as it actually moves from one to the other, and because of this memory you see the movement of the hand, even though for most of the time the hand is stationary. (There are echoes of Zeno's arrow paradox here; see p12.)

WHAT DO WE MEAN BY NOW?

This is one of those recurring fundamental questions. In his philosophical essay *The dimension of the present moment* Czech immunologist and poet Miroslav Holub wrote: "The fact that I cannot imagine the present moment has always worried me... As a matter of fact, I can imagine eternity much better, particularly when looking up at the sky or the ceiling of a waiting room." He goes on to say, however, that experimental psychologists have shown the "present moment" to be about 3 seconds long, and that this is the natural length of a line of verse or a segment of speech.

William James

Born in New York City in 1842, William James, elder brother of the novelist Henry James, became a professor of philosophy at Harvard University, and one of the founding fathers of the discipline of psychology. He wrote extensively about the subject, notably in *The Principles of Psychology* (1890), which includes a substantial chapter on "The perception of time".

He begins by pointing out that to live a normal life we need a notion of "pastness". If consciousness were like a string of beads – unconnected sensations and images – "we should never have any knowledge except that of the present instant. The moment each of our sensations ceased it would be gone for ever. And we should be as if we had never been." So we should be unable to acquire knowledge or experience. This notion of pastness is the basis of memory and of history. On the other hand to appreciate that two events have happened in succession we must at the present think of those two events, even though at least one of them is definitely in the past. We can never think of two sequential events while both are happening, since Now one of them must be over.

BELOW: People who witness a gunfight hear guns firing and see bodies falling in a confused rapid sequence. Only afterwards do their brains work out what must have happened, and the order of events. This made-up story will form the basis of their memories, but three different observers may tell three different stories about the same gunfight.

Any knowledge or feeling of the present is always coloured by the immediate past, and to some extent by the future. So for example if you are thinking of the sequence A B C D E F G then the next one will be B C D E F G H, and the following one C D E F G H I, "the lingerings of the past drop successively away, and the incomings of the future make up the loss." These lingerings and incomings are the basis of memory and expectation, which together with the present make up the stream of consciousness.

At a pedantic level, you cannot perceive the precise present, since information in the form of light, sound, and so on takes time to reach you, and your brain takes time to process the information. Therefore you can experience only the

BELOW: In James's "specious present", our eyes do not follow a bird's flight over a series of distinct points in space. Rather, we perceive the *movement* of flight.

immediate past. As soon as it has been processed, this information must be consigned to memory, since fresh information is coming in, and you cannot simultaneously process successive packets of conflicting information. So you directly experience the immediate past, but you cannot directly experience either the future or the distant past.

The specious present

The French philosopher René Descartes thought of time as a series of instantaneous Nows, while Isaac Newton believed in a knife-edge separating past and future. William James (following psychologist E R Clay) described instead a "specious present", lasting about 12 seconds, which is what we feel in the flow of experience. All the notes of a bar of a song, for example, seem to lie in the present, even though they did not all happen at the same instant and must all have been in the immediate past. You have to be aware of all the words of this sentence as a whole, even though you cannot read them all simultaneously.

"Now", therefore, according to James, is "no knife-edge, but a saddle-back, with a certain breadth of its own on which we sit perched, and from which we look in two directions into time" – backwards into the remembered past, forwards into the expected future. This specious present may correspond to the length of short-term memory, or to the duration of whatever is perceived as instantaneous. It must also include movement. You can see a bird flying; you do not note that at one instant it has moved from point A to point B; you actually directly perceive the movement, just as you perceive the movement of the second-hand of a watch and of the arrow in flight.

The specious present makes sense of common experience. If you are deeply immersed in reading a book, or writing a sentence, you may suddenly become aware that a clock has struck. What is more, you may often be able to say how many strokes there were. You were not listening for the strike, nor consciously aware of it when it happened, and yet immediately afterwards you were able to say what hour it struck. The information must have come into that specious present. In the same way you may, while deeply engrossed, become aware that the telephone has been ringing for some time before you come to notice it.

Duration

In *An Introduction to Metaphysics*, James's colleague and friend, French philosopher Henri Bergson, described an image of the Duration (the inner life of a person): two spools, one unrolling to represent the continuous flow of ageing as you feel yourself moving towards the end of

LEFT: Where is Now? It vanishes like a snowflake in the palm of our hand and "melts in our grasp", according to James.

your life span, the other a thread rolling into a ball to represent the continuous growth of memory as your past follows you. No two moments are identical, for one will always contain the memory left it by the other. Bergson came to the conclusion that time is mobile, so that as soon as you attempted to measure a moment, it would be gone. What you measure is an immobile, complete line, whereas time is mobile and always incomplete.

Meanwhile in relation to the Now, James goes on to say, "Let any one try, I will not say to arrest, but to notice or attend to, the present moment of time. One of the most baffling experiences occurs. Where is it, this present? It has melted in our grasp, fled ere we could touch it, gone in the instant of becoming."

Now again

Here again is the mystery of Now, which worried Aristotle, Newton, and Leibniz (see pp12, 16–18), and will probably never go away. Are you thinking about it Now?

This is interesting in the light of Buddhist-style meditation, when the teacher frequently tells the pupils to "be in the present", to "focus on what is now", to "be mindful", or "let thoughts go away and keep your mind open to experience". How can a meditator experience the Now if William James could not?

TRAIN OF THOUGHT

A train of thought can be exceedingly fast. In an instant your mind can leap from the smell of a succulent dish to a holiday long ago to a long-lost lover to a railway carriage to a favourite book to a curious patch on the ceiling. The experiences are so fleeting they tumble over one another, and often the train is hard to retrace. Sigmund Freud asserted that the unconscious is timeless; in dreams long sequences of events can happen in seconds, and events may happen in random order.

RIGHT: Although the content of the specious present is in flux, with new events coming in as quickly as they fade out, James argued that the specious present itself "stands permanent, like the rainbow on the waterfall, with its own quality unchanged by the events that stream through it."

In the 1970s, before the discovery of modern drugs, brain surgery was much more common than it is today. Many epileptic patients had their skulls opened, so that electrodes could be placed on their brains while they were wide awake.

Time and consciousness

AMERICAN NEUROSCIENTIST BENJAMIN LIBET took the opportunity to ask some of these patients whether they would assist in some extraordinary experiments on consciousness.

Libet stimulated the sensory area of these patients' brains with rapid trains of tiny impulses that, if they were strong enough, felt to the patient a bit like a touch on the arm. But he found an odd thing. Even if the pulses were strong enough to induce a sensation, the trains of impulses had to carry on for at least half a second, or the patient would report feeling nothing at all.

This remarkable discovery, which became known as "Libet's half-second delay", implies that consciousness lags half a second behind the events of the outside world. But this seems incomprehensible. We know that people react much more quickly to what's going on around them; if you accidentally touch a hotplate you can pull your finger back in about a fifth of a second. So do you react first, but only become conscious of the event a bit later?

From these and other experiments Libet concluded that consciousness does indeed take time to build up, but then is "backdated" to the correct time. So when someone taps your arm, what happens is that your brain first reacts, and can make you snatch your arm away if necessary, then waits to see whether the activity in sensory cortex goes on for at least half a second. If it does, your brain backdates the feeling of being tapped to seem as though it came at the time of the tap.

Things are even stranger when it comes to conscious, deliberate actions. In the 1980s Libet did new experiments in which he asked people to flex their wrists at a time of their own choosing – using their own free will. He then measured three things – the time at which their wrist moved, the time at which brain activity started up in their motor cortex (the part of the brain that organizes and sets off movements), and the time at which they consciously decided to act. To his surprise he discovered that the "decision" to act came a

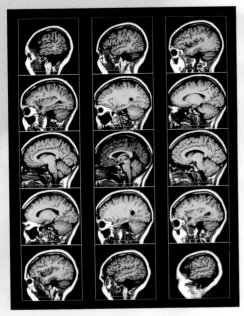

BELOW: An fMRI scan of a child's brain. Advances in neuroscience have revealed the time delay inherent in consciousness.

full half second after the brain activity started. Once again it seemed that consciousness was lagging behind the real action. Free will, some contended, must therefore be an illusion. Neuroscientists and philosophers have been arguing about the implications of Libet's delay ever since.

On the other hand if one of the subjects changed his or her mind and aborted just before actually moving the wrist, the "decision" seemed to precede the brain activity. In other words, according to psychologist and punster Richard Gregory, "We may not have free will, but we do have free won't."

And lurking in that great scientific mystery the nature of consciousness are various other peculiar time effects that cause havoc with our intuitions. Imagine once more that you suddenly notice a clock is striking. You have only just become conscious of the sound, and yet now you can count backwards and know that you missed three strikes, listened to two more, and so it must be five o'clock. Were you conscious of the missing three? No, because you felt as though you became aware of the sound on strike four. On the other hand, yes, because you can remember all the strikes in sequence. William James pointed out this peculiarity long ago (see p23), but it has recently been taken up again by philosophers and neuroscientists.

Intuitively we assume that either I was, or I was not, conscious of something that happens. Not so, says American philosopher, Daniel Dennett, author of *Consciousness Explained* (1991). He claims that most of our intuitions about consciousness are simply wrong, and that in fact there is no answer to this conundrum.

RIGHT: We can become conscious of a clock striking, and count backwards to "update" our minds with the number of chimes.

Those who practise meditation suggest that time is an illusion, and that the more we focus on the present moment the less we experience the Now – which is all that there is. To quote the German spiritual teacher Eckhart Tolle (*The Power of Now*, 1997), "The eternal present is the space within which your whole life unfolds, the one factor that remains constant. Life is now."

TIME AND BUDDHISM

THERE NEVER WAS A TIME WHEN YOUR LIFE WAS NOT NOW, argues Tolle. Nor will there be. "Have you ever experienced, done, thought, or felt anything outside the Now? Do you think you ever will? Is it possible for anything to happen or be outside the Now? The answer is obvious, is it not?"

"Nothing ever happened in the past; it happened in the Now. Nothing will ever happen in the future; it will happen in the Now."

In his book *Catching a Feather on a Fan* (1991) British psychologist John Crook writes, "As time passes, you witness the passage of thoughts. As thought succeeds thought, you experience the passage of time. In your practice it is important to make every thought the present moment. If you make yourself one with the moment, you stop the thought." Without thought, he writes, time simply becomes a continuous present. Then you learn to experience what it means to be one with the moment: "Everything is continuously fresh, like the water of a spring endlessly bubbling up into the open air ... As one ancient master has said 'One thought for a thousand years.' Yet in this thousand years, there are no thoughts. There is simply a continuous unbroken newness."

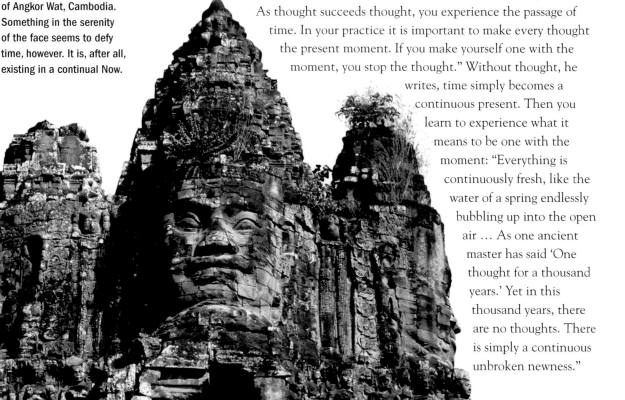

BELOW: Time has ravaged this Buddha at the temple of Angkor Wat, Cambodia. Something in the serenity of the face seems to defy time, however. It is, after all, existing in a continual Now.

FLOW

Hungarian-American psychologist Mihaly Csikszentmihalyi has described a mind state called "flow", which he observed in artists, athletes, chess-players, and mountaineers, among others. An artist becomes so completely involved in her work that she does not notice the passage of time, forgets about meals and sleep, and thinks only about her paint, her brushes, and her canvas. Indeed for her there is no separation between these elements: nothing but the process of painting. This is not because she wants to finish the work and perhaps sell it, but just because she wants to paint. Mountain climbers, he says, can become so transfixed by the process that they are no longer climbing to reach the top; instead they are going for the top in order to climb. This is what he calls "flow". Not surprisingly, meditation and mindfulness help people to achieve flow.

The 13th-century Japanese Zen master Eihei Dogen asserted in the *Shobogenzo* that a day consists of 6,400,099,180 moments (which makes a moment about 1/74,000 of a second). In *Each Moment is the Universe* (2007), Zen monk Dainin Katagiri writes that these moments are so fleeting that our rational minds are too slow to keep up with them, and so we experience a gap between us and the reality of time, which is why we feel we can't keep up, and why we suffer. According to Buddhist teaching, everything exists together in a moment.

The past has already gone, says Katagiri; so it does not exist. The future has not yet come; so it also does not exist. So the past and the future are nothing, no-time. Then is the present all that exists? No, even though there is a present, strictly speaking the present is nothing, because in a moment it is gone. So the present is also nothing, no-time, no-present, no form of the present.

But that nothingness is very important. The real present is not exactly what you believe the present to be. In everyday life we constantly create some idea of what the human world is because we are always thinking

BELOW: An 1811 edition of the *Shobogenzo*, literally "Treasury of the True Dharma Eye". The name can also refer to the essence and realization of the Buddha's teaching.

Nothing ever happened in the past, it happened in the *n o w* Nothing will ever happen in the future; it will happen in the Now (Eckhard Tolle)

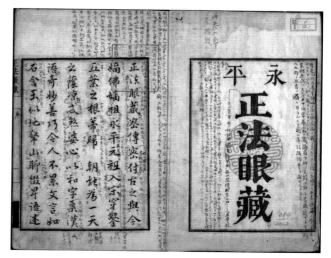

about how things were in the past or how things will be in the future. The real present is the full aliveness that exists at the pivot of nothingness. Be present, Katagiri says, from moment to moment, right in the middle of the real stream of time.

In the *Shobogenzo* Dogen says that when you swim on the surface of the sea, your foot touches the bottom. The surface is the "normal" human world in the stream of time, the world we create with our imagination, memory, and hopes, while the bottom is the reality of human life. So the surface life is constantly changing, but the bottom is the firm reality, and we always swim with one foot on the bottom.

Be present, Katagiri says, from moment to moment, right in the middle of the stream of time

The distracting present

A paradox of Zen is that the present is what we all experience all the time; so why is meditation and trying to focus on the present moment so difficult? And it certainly is difficult; I have tried it. I have sat and looked at a wall for several hours, on several days, and tried in vain to get rid of all the thoughts that besieged my brain; indeed my brain wound up in turmoil. Afterwards the teacher said, "Well, there's only you and the wall – and there's nothing wrong with the wall."

Roshi Philip Kapleau, in his book *The Three Pillars of Zen* (1965), tells a story about one of his teachers Yasutani-roshi answering a question put to him by a bunch of American students: If, as we have been led to believe, … the past and the future are unreal, is one not free to live as one likes in the present, unconcerned about the past and indifferent to the future? In reply, Yasutani-roshi made a dot on the blackboard and said that this isolated dot represented their conception of "here and now". To show that their idea was hopelessly incomplete, he put another dot on the board, and drew through it a horizontal line and a vertical line. The horizontal line stood for time from the limitless past to the endless future, and the vertical one for limitless space. The "present moment" of the enlightened man, who stands at this intersection, he explained, embraces all these dimensions of time and space. The realization that you are the focus of past and future brings a sense of responsibility to your family, to society as a whole, to those who came before, and to those who will come afterwards.

OPPOSITE: For those living in the moment, Now is neither a knife-edge nor a saddle, but the entire universe, unchanging, the space where everything happens.

BELOW: One of the colossal statues of the Buddha in Japan. Zen buddhism emphasizes direct realization through mediation and *dharma* (universal truth).

No-one has yet discovered a particular sense organ for perceiving time directly, but it seems that it may be a cognitive process – a matter of concentrating and comparison with other events. There is more about this is chapter 2, but here I discuss some psychological observations.

ABOVE: How long does it take your nerves to tell your brain to move your hand when a stinging insect lands on it? Reaction times get longer with age.

TIME INTERVALS

WE ARE FUNDAMENTALLY AWARE OF TIME intervals of only a few seconds. When we think about longer and more useful intervals – minutes, hours, and days – we have to conjure up symbols and mentally add them together. According to William James, we cannot estimate the duration of an event by noting the time of the beginning and of the end and then subtracting one from the other. Rather, we "feel the interval of time as a whole, with the two ends embedded in it." Observing a rapid sequence of events, we may well be able to distinguish individual events, but be unable to be sure about the order in which they happened, which is one reason why eyewitness accounts of accidents often differ, even though the witnesses have actually seen the same things happening.

James measured the shortest interval that can be perceived by eye by arranging for two sparks to appear in rapid succession. Using a simple electrical apparatus with no modern electronic timing system he found that the shortest perceivable interval between the sparks is about 44 milliseconds, or 1/23 second. Today scientists measure the "flicker fusion frequency", or FFF, by showing you a light that flashes on-off-on-off, and gradually increasing the rate of flashing until you are no longer aware that the light ever goes off. That rate is your FFF; it is generally around 60 flashes per second for humans.

Some old fluorescent light tubes flicker, because they go bright and dim with the frequency of the electricity mains, which is 60 cycles per second (Hz) in the United States, and 50 Hz in much of the rest of the world. Modern tubes have electronic

REACTION TIMES AND MENTAL CHRONOMETRY

Your reaction time is the delay between your receiving a sensory stimulus and your response. In this simple case it is a measure of how long it takes for nerves to convey the information to the brain and how long the nerves and muscles take to respond. Such reaction times are shortest for young adults, and increase with age. Mine, according to a couple of interactive websites, is about 0.28 seconds.

Imagine you are asked to press a button as soon as a light comes on. You do this ten times and have an average reaction time of 0.26 seconds. Next you are faced with two lights, and asked to press the button when the left light comes on, but not

if the right one comes on. This will take longer, because you have to decide which bulb has come on, and then press. The delay is a measure of how long it takes you to decide between the lights. Then you might have two buttons and two lights, and be asked to press the left button when the left light comes on and the right button when the right light comes on. Again this will take longer because of the extra processing needed.

Many recent studies in mental chronometry have helped to elucidate not only how we process data and how memory works, but also, with the help of EEG and fMRI scanning, where in the brain each type of processing takes place.

ballasts and phosphor coatings which effectively prevent this flicker. Old movies, projected at 24 frames per second, show flickering images, while modern television pictures are generally shown at an effective rate of 60 Hz, so that the motion appears continuous.

Flies have FFFs as high as 200 Hz. Birds have been shown to have FFFs up to at least 100 Hz – considerably faster than humans – which is why they can fly fast through woodland without crashing into branches. Unfortunately car drivers are not equipped with high FFFs, which is one reason why there are so many crashes on the roads.

Hearing

The sense of hearing seems to be more acute for time intervals. James found that the shortest perceptible interval between two clicks is just over 2 milliseconds, or 1/500 of a second, which is about ten times shorter than the shortest visible interval. Two clicks closer together than this are heard as a single louder click.

The longest cluster of clicks or notes that could be heard as a single group was around 40, in groups of 5 or 8, played over 12 seconds. This seems to be the longest "filled duration of which we can be both distinctly and immediately aware," and corresponds (along with "a vaguely vanishing backward and forward fringe") to James's "specious present".

James also points out that an interval between two sounds seems longer if nothing happens in between than if the gap is filled with other sounds, just as any period filled with action seems to pass much more rapidly than one in which nothing happens. Short periods of silence seem particularly long when they are deliberately included in the middle of a piece of music or a speech; the listener is made to want the next phrase. This is why the dramatic pause can be so powerful.

ABOVE: Projected at 24 frames per second, movies from the silent era appear to flicker.

Humans can perceive up to 60 on-off FLASHES per second – our "flicker fusion frequency". Flies can perceive up to 200

Humans can hear two sounds separately even if they are only about 2 milliseconds apart

Human hearing systems are extraordinarily adept at distinguishing time intervals too short for conscious perception; this is how we can tell where a sound is coming from.

Imagine you are standing in a street, or on a beach, and you hear a shout; you know at once where the shout has come from. You hear only a single shout, but the sound reaches your ears at slightly different times, and you can unconsciously pick up the different arrival times.

Suppose you are facing north, and the shout comes from exactly north (in front of you), or south (exactly behind you), then the sound will reach both ears at the same instant. But suppose it comes from your left – from the west, or the port beam, in nautical terms. The sound waves in the air reach your left ear first, and have to travel a further 25cm (10in) or so, depending on the size of your head, to reach your right ear. Sound travels through air at about 340m/s (1,100ft/s) or 340mm/ms (13in/ms); so it will take about 0.74ms to get round your head. This is a much shorter interval than you can consciously notice (which is about 2ms), so you perceive the sound as a single shout, and yet your brain can distinguish the two arrival times and work out that the sound comes from the left. But the brain is much more cunning than this. If the sound comes from the north-west, then the sound waves will take perhaps only 0.4ms longer to reach your right ear, and yet you can still tell immediately where the sound has come from. Indeed your hearing can probably allow your brain to calculate the origin of a sound to within about 20°, even though you cannot consciously perceive the delay in its arrival at your two ears.

BELOW: In music, as in theatre, a deliberate pause can be used to dramatic and powerful effect.

Filling the gaps

In practice, estimating the length of a featureless interval is extremely difficult, as we tend to fill such a gap with thoughts – thoughts that the Buddhist meditators try to drop. If you want to estimate the time you can try counting ("elephant 1, elephant 2, elephant 3…" for seconds) or attend to your own pulse or rate of breathing. And yet some people can accurately announce the time of day or night without consulting a clock, and others can wake regularly at a prearranged time without any kind of physical alarm, even if that prearranged time varies from day to day.

We all experience the apparent slowing down and speeding up of time – it drags in a boring situation and races by in an exciting

one. Einstein remarked "When a man sits with a pretty girl for an hour, it seems like a minute. But let him sit on a hot stove for a minute and it's longer than any hour. That's relativity."

Recent research has shown that when subjects are asked in advance to estimate the length of a time interval they tend to overestimate if nothing is happening during the interval; in other words they perceive empty intervals to be longer than filled ones, even though they are actually the same length. In another study, 80 children aged between 7 and 9 were asked to estimate the times (a few seconds) for which light bulbs were switched on. Their estimates were longer when they had nothing else to do but concentrate on the lights, and shorter when they were distracted by other tasks.

All sorts of experiments have been done to investigate factors that affect our perception of time. For example people were asked to say whether a particular image was shown for a short time (about half a second) or a long time (about one and a half seconds). They consistently underestimated the time when the image was of food, and even more so when it was food that they disliked; in other words they were so busy thinking about how much they disliked the food that they forgot to think about how long the picture was in view.

Time flies when you are busy – people often underestimate time when they are concentrating fiercely on tasks that need attention, probably because their attention is diverted from keeping track of the time, so that it slips by unnoticed.

ABOVE: For bored passengers delayed in a departure lounge, time may seem to dilate, stretching because nothing is happening to fill it.

A CHANGE OF PLACE AND PACE

In his book *Making time* (2007), Steve Taylor describes the reaction of a student who had just returned home after her first term at university, where she had encountered a myriad of new people and experiences: "Time was so slow that by the time I went home at the end of the first term it seemed like about two years. So much seemed to have happened in the three months. I felt like I'd been away for so long that I was surprised that everything was the same at home."

Conversely people overestimate time – think it is passing slowly – when they are bored, anxious, nervous, or just waiting. Perhaps the worst is a delay in anticipation of an event – the last half-hour of the school day can seem like hours as you wait for the final bell, and my mother used to tell me that a watched pot never boils. In a sense Aristotle had a point when he said that time is all about change and events; we find it hard to perceive the flow of time if nothing is happening.

In retrospect, however, the reverse is often true: a stretch full of action and excitement seems to have taken a long time, while a tedious period of boredom is condensed to a brief moment. As James points out, "a week of travel and sightseeing may subtend an angle more like three weeks in the memory; and a month of sickness hardly yields more memories than a day."

Older people often perceive time as passing more quickly than young people. This may be because for young people all experiences and therefore all memories are fresh and new and loom large on the horizon, while for older people most experiences and memories are repeated; the feelings are therefore less arousing and interesting. As a result the memories are condensed and seem to have occupied less time. In an experiment, however, when a group of old people

Time seems to **SPEED UP** as you age, because experiences are repeated and memories **CONDENSE,** seeming to take up less time

(aged 60–80) were asked to say when 3 minutes had passed, their estimate on average was 3 minutes 40 seconds, while a young group (aged 19–24) estimated 3 minutes 3 seconds.

Psychoactive drugs often alter your perception of time. Stimulants such as caffeine, nicotine, ecstasy, and cocaine, may lead a person to overestimate time intervals, while depressants such as alcohol and barbiturates may appear to shorten them. Meanwhile LSD, psilocybin mushrooms, and mescaline can cause time estimation to be scrambled: time may appear to speed up, slow down, stop, or run backwards.

Animals are quite good at estimating time. Experiments have been done with many species, including rats, cats, dogs, and pigeons. For example when they hear a sound that lasts a particular length of time then they may get a food reward if they press and hold a lever for two seconds. Or they have to wait 30 seconds and then press a lever to get the reward. Like humans, animals can be distracted by other activities, and their perception of time can be altered by psychoactive drugs.

BELOW: Psychoactive drugs such as ecstasy can make time condense, with several hours seeming to squeeze into one.

People asked to perform a task in a limited time are under time stress. This includes participants in television game shows, doctors in accident and emergency departments, pilots about to land aircraft, and even authors trying to write to deadlines.

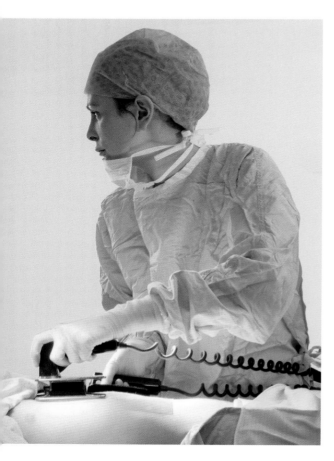

ABOVE: Your whole life may flash before your eyes during an NDE. With improvements in cardiac resuscitation techniques, the number of reported NDEs is rising.

Events can appear to **u n f o l d** in slow motion during an emergency

TIME STRESS

THE PROBLEM SEEMS TO BE that they spend part of their time and energy worrying about how much time is left for the task and whether they are keeping on schedule. This means they have less concentration available for the task itself. Under these conditions people can easily make mistakes – or bad decisions – which is one reason why those occupations tend to offer good rewards for those who can cope with the stress and work efficiently up to a deadline.

Among the problems caused by such time stress for those who lack the skills to cope are a tendency to process first whatever information seems to be important, to forget other pieces of information, to focus especially on negative information, and to make wrong judgements, later fiercely defending such mistakes, often on spurious grounds.

Life review

People facing imminent death, trapped under water and about to drown, for example, report seeing their whole lives flash before their eyes. This is not an experience that I wish to test, but it has been widely reported and studied. About five per cent of those who have near-death experiences (NDEs) report a life review, and it is most common among people who nearly drown.

Apparently a whole jumble of images from the past appears suddenly, almost in an explosion, either in chronological order or all at once. They are seen in great detail: often the images are described as "extra-real" or "three-dimensional". The whole process seems to vary in duration from instantaneous to a few seconds. Often these life reviews include incidents of cruelty inflicted on others by the person involved; hitting someone in the school playground, screaming abuse from a window, or crashing into someone else's car. But at the same time there is an immensely powerful

IS TIME ALL SUBJECTIVE?

In life reviews, free fall, and prison cells, time seems to be elastic - stretching and shrinking according to the psychological state of the individual. Does this mean that time depends on us, and cannot really be independent and absolute, as Plato and Newton believed? Or is it just that we are not always good observers, or our senses for estimating time are too unreliable?

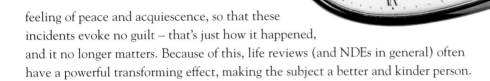

feeling of peace and acquiescence, so that these incidents evoke no guilt – that's just how it happened, and it no longer matters. Because of this, life reviews (and NDEs in general) often have a powerful transforming effect, making the subject a better and kinder person.

Emergency slow motion

Almost the opposite effect is observed during an emergency. Everything seems to slow down and actions happen in slow motion. The emergency may be trivial – I remember in a smart French restaurant seeing my wife make an extravagant gesture and knock over a full glass of red wine; it seemed to take minutes for the glass to fall and the wine to spread into a crimson lake on the crisp white linen tablecloth. More often this is reported in serious life-and-death incidents, often during road accidents.

Even this has been investigated, by Professor David Eagleman, Director of Baylor College of Medicine's Laboratory for Perception and Action in Texas, and his intrepid graduate student Chess Stetson, who became the first fall guy. Eagleman built a 50m- (165ft-) high tower from which volunteers could fall backwards into a net, and so be in free fall for three seconds. During this terrifying fall they reported feeling that they were moving in slow motion, and when they were asked afterwards to re-create the fall in their minds they estimated it had taken between four and six seconds rather than three.

To find out whether they could "see" more quickly during the fall, Dr Eagleman designed and built a wristwatch (a "perceptual chronometer") on which was displayed a random number flickering with its negative image just faster than was normally discernible (about ten times per second). In the laboratory all the volunteers could see was a blur, but would they be able to see the separate images if their sense of time was speeded up, so that they saw the world in slow motion? The answer was no; they could not.

ABOVE: Even minor emergencies, such as knocking over a glass of wine, may seem to happen in slow motion.

ABOVE: Time seems to slow dramatically when you are in free fall.

Eagleman reckons that in such an emergency situation a small part of the brain called the amygdala "kicks into high gear", and memories are laid down by a secondary memory system, where they seem to stick better. Therefore you lay down more memories during the emergency, and as a result the event seems to have taken longer than it really did. This may help to explain why time seems to speed up for old people; their memories are compressed and therefore seem to have taken up less time.

THE AMYGDALA

The amygdala, a pair of almond-shaped structures deep in the base of the brain, is responsible for processing and remembering emotional reactions. If you are afraid of snakes, your amygdala will have stored this fear. You may freeze; your heart may beat faster; you may breathe rapidly. The amygdala is to blame. It is also involved in your sense of smell, and in the slow process of laying down long-term memories.

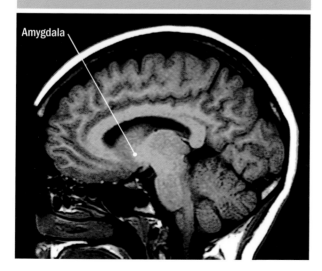

Amygdala

Time in isolation

From inside a prison cell, time seems different. John McCarthy was kidnapped in Beirut in 1986 and kept hostage for five years, during which time he was utterly cut off from family and friends. In his book *Some Other Rainbow* he wrote, "I was to be in this solitary cell for less than three months, but after the first two or three weeks it felt as if I had slipped into a different time-scale. Days passed without any variation… In those first terrible weeks… the date and the number of days of my captivity was the only information I had. Counting the days became an obsession."

Terry Waite was seized in Beirut in 1987, and spent nearly five years in captivity, most of it in isolation. He wrote memorably of the experience

For prisoners, the hours may seem to drag, but the years   F L Y

in *Taken On Trust*. "How many days have passed? Five? Six? It's so easy to lose track… The days are so long. Apart from exercising my body and my mind, there is nothing to do… the whole day lies ahead like a vast unexplored sea… The days are never-ending…"

Brian Keenan, likewise locked up in Beirut in 1985 and on his own for years in a tiny cell, wrote vividly of the experience in his book, *An Evil Cradling*. "Days now passed in an excruciating boredom in which the mind ran hither and thither looking for a place, an idea, a memory of the past in which to hide and absent itself from this tiny cell… My mind now moves into strange abstractions. The idea, the concept of time enthrals me. I build a complicated and involved structure which redefines what time is. Time is different now. Its flux and pattern is new, seeming so clear, so precise, so deeply understood yet inexplicable. I am calm and quiet… Time was taken from me. How long, how long would I be here?"

The first experiments to investigate how people perceive time without external stimuli were done in caves. In the summer of 1938 psychologist Nathaniel Kleitman and his student Bruce Richardson spent a total of 32 days in a deep cave in Kentucky, cut off from any normal time-clue such as daylight, and tried to adjust their normal "day" to 28 hours, rather than 24. The plan was to sleep for nine hours and stay awake for 19, which would mean that six days for them would correspond to seven above ground. The results were inconclusive: Richardson, aged 25, was able to adapt to the longer day, but Kleitman, aged 43, was not.

This early experiment was a deliberate attempt to change the "normal" day length, but there is a more fundamental question: what, from the human perspective, is the "normal" length of the day?

In July 1962 French caver Michel Siffre, who seemed to enjoy spending time on his own underground, lived for two months in the Scarasson Cave in the French Alps. He used a one-way telephone to tell his team on the surface when he woke up, when he ate, and when he went to sleep. They in turn controlled the lights in the cave, switching them on when he woke and off when he went to bed, but they provided no information about the actual time. When he came out of the cave he thought the date was 20 August, but it was actually 17 September; so his estimation of time was way out. Several further investigations of this kind have been carried out, and the nature of the effect of isolation on biological clocks is discussed further in chapter 2.

BELOW: Even for prisoners on short sentences, time can become an obsession.

Conceptions of time are heavily dependent on culture and environment. Intrepid Polish traveller Ryszard Kapuscinski wrote vividly about African time in his book *Shadow of the Sun,* published in English translation in 2001: "The European and the African have an entirely different concept of time..."

THE PACE OF LIFE

IN THE EUROPEAN VIEW, according to Kapuscinski, time exists outside man, exists objectively, and has measurable and linear characteristics. "Africans apprehend time differently. For them, it is a much looser concept, more open, elastic, subjective. It is man who influences time, its shape, course, and rhythm… Time appears as a result of our actions, and vanishes when we neglect or ignore it."

The climate, and especially the temperature, must have quite a lot to do with this attitude. Outside Africa, the pace of life is said to be fastest in Switzerland

BELOW: In much of the developing world, the pace of life is determined by wet and dry seasons, planting, and the care of livestock.

and Germany, and slowest in Indonesia, Kenya, and Mexico. People who live in "rapid" zones are often scathing about the slowness and unpunctuality of the others: the Spanish word *mañana*, meaning "tomorrow", or more strictly "in the morning", is often perceived as a delaying tactic. When I moved to Devon in south-west England I learned a new word, "dreckly" (directly), as in "We'll deliver it dreckly", which actually means "at some indeterminate time in the future". In other words, dreckly is like *mañana* but without the same sense of urgency.

ABOVE: The top speed of cars has quadrupled since the 1920s. In the air, low-cost jet travel has enabled business travellers to cross multiple time zones and be home the same day. Instant communication by e-mail and cell phone has made everyone available 24 hours a day, and thereby hugely increased the pace of life, and the pressure of time.

The idea that the pace of life is continually increasing is common, and not at all new. In June 1825 the German poet Goethe wrote to his friend Carl Friedrich Zelter, "Everything… is now ultra, everything transcends… Young people are stimulated too soon, and swept along in the whirlpool of time; wealth and speed are what the world admires and what it strives for. All sorts of communicative facility are what the civilized world aims at in outpacing itself."

In western cultures there is ever-increasing pressure on everyone to do more and do it more quickly. Time is seen as a precious resource, not to be wasted. Many people feel inadequate, and unable to complete all the tasks in front of them; they complain of "time-poverty". This is brought on partly by the ever-increasing amount of technology that surrounds us. Much of it was initially designed to take the drudgery out of life and give us more leisure time, but in fact it has had the opposite effect; it has made us think we can do more work.

Outside Africa, the pace of life is *fastest* in Switzerland and Germany

Increasing amounts of time and increasing numbers of courses and books are devoted to "time management" – to organizing our lives in order to get all those things done. Inevitably the feedback is not wholly positive – by getting more done we increase not only productivity but also fatigue, the chances of repetitive strain injury, and the psychological pressure.

Bizarrely, the way people think about time seems to depend partly on their temperament. Imagine you have been told that a meeting originally scheduled for Wednesday has been moved forward two days. Experiments have shown that people with an angry disposition are likely to feel themselves moving through time, and will assume it has been moved to Friday, while peaceful people tend to feel time is moving towards them, and will expect it on Monday.

The drive to increase industrial productivity using scientific management took off in the early 20th century with the work of Frederick Winslow Taylor and Frank and Lillian Gilbreth.

Time study

ABOVE: Time-study expertise came to the fore during World War II, when vast numbers of ships and aircraft had to be produced in the shortest time possible.

TAYLOR WAS BORN IN PHILADELPHIA IN 1856, became an engineer, studied the process of work, and became the "father of scientific management", sometimes known as Taylorism. His method was to break any job down into its component parts and time each with a stop-watch. Taylor was the original "time-study man". He proposed that managers should work out the scientifically best method for doing any job, and then train the workers to do it that way. Managers should prevent workers from being interrupted, and should provide wage incentives to increase productivity. The life and outlook of a time-study man was caricatured and immortalized in the song "I'm a time-study man" in the musical *The Pajama Game* (1954).

Frank Bunker Gilbreth (1868–1924) began his working life as a bricklayer, and progressed via inventing to become a management engineer. He married Lillian Moller in 1904, and together they not only had 12 children but developed the ideas of time and motion. They used a new-fangled motion-picture camera to record and analyze the actions of workers.

The classic example of scientific management to increase productivity was in bricklaying. Having been a bricklayer himself, Gilbreth carefully studied a bricklayer at work, and noted that he performed 18 separate actions in laying bricks. By analyzing the process he was able to reduce the number of necessary actions to five, which greatly increased the productivity, and also reduced the fatigue of the bricklayer.

The Gilbreths invented the term "therblig" (which is almost Gilbreth backwards) to describe each of all 17 possible movements needed to carry out a task (Search, Find, Choose, Grasp, Hold, etc.). They went on to develop "time and motion" ideas in the 1920s, asserting that there was "one best way" to carry out any process. Frank Gilbreth later suggested that during operations, surgeons should be assisted by "Caddies" – surgical nurses, who would hand instruments to the surgeon when they were needed. He also suggested that army recruits should be taught to take their weapons to pieces and reassemble them while blindfolded, so that they would always be able to do it in the dark.

During the subsequent decades the ideas of scientific management evolved into quality control, operational research, and cybernetics. Meanwhile the person with the clip-board and stopwatch is now seen less on the factory floor and more on the sports or athletic field, where time and motion is all.

BELOW: Improving technology enables operators to handle ever-increasing volumes of work. The paradox of this is that the more efficiently work is done, the less time people seem to have.

All these ideas in philosophy and psychology, and the perception and use of time, do not seem to answer that fundamental question of whether time is absolute, or whether it is determined by events. Perhaps we can get closer to an answer by considering whether time had a beginning, and when it will have an end.

THE BEGINNING OF TIME

ABOVE: God creates time, from the ceiling of the Sistine Chapel (1508–12, by Michelangelo).

WAS THERE A BEGINNING OF TIME? Cosmologists have answers to this question, which are discussed in chapter 5, but there are many other answers in the creation myths of civilizations around the world.

Early Christians argued for centuries about whether God created everything in one go, or whether He had to set up a matrix of space and time into which He could place all the real stuff. Before the Sun had been created, for example, there could have been no shadows, and therefore no sundials to measure time. This raises another philosophical question: can there be time without measurement? The Fourth Council of the Lateran in 1215, convoked by Pope Innocent III, was attended by 71 patriarchs and metropolitan bishops, and 412 bishops. They considered the fifth crusade, the measures to be taken against heretics, and 71 decrees; but most important they decided that God created, in a single event, "all things spiritual and corporeal, angelic and mundane".

There remained the question of what time the universe was created – surely it must have been at some specific time – and various beliefs grew up among the cultures around the Mediterranean about whether it was spring or summer, and

the position of the Sun. Further afield, each culture had its own creation myth. In Southern India the god Shiva carries in one of his four hands a dumaru, or drum of creation, since sound was the first thing that happened in the universe. In Mesopotamia the spring equinox and New Year were celebrated in honour of Marduk, the god who acquired the powers of Ea and Enlil, and created the world by defeating his enemy Tiamat. His heroic life story is told in *Enuma Elish*, a creation story of a thousand lines inscribed on seven clay tablets.

The wheel of time

In the Judaeo-Christian tradition time began with God's creation of the universe, and always moves forward in one direction; but many other cultures, both past and present, believe in cyclical time – time that goes round and round and repeats itself. In linear time each event occurs only once, which means that if you want to make progress you must grab the moment, but if time is cyclical then everything and every opportunity will come round again some time. Fate and the wheel of fortune will decide your destiny.

According to Professor Giorgio de Santillana at Massachusetts Institute of Technology, more than 30 ancient cultures believed that history consisted of a repeating sequence of dark and golden ages, which may have been linked to the precession of the equinoxes (see p54). The Incas, their descendants the Q'ero Indians in Peru, the Maya, the Hopi Indians in Arizona and New Mexico, and other American Indian tribes had a concept of a wheel of time, a never-ending cycle.

Aztec and Maya cycles

The Mayan civilization was first established in Mesoamerica some 4000 years ago, and reached its heyday around 1500 years ago. The Maya were able to record events in a linear sequence – with their "Long-count" system they were able to record individual days over a period of 7885 years – but their concept of time revolved around cycles. They saw the obvious cycles of days and nights, waxing and waning

The Mayan great cycle of time began in 3113 BC and is destined to end on **12 December 2012**

BELOW: Temple 1, a pyramid at the ruined Mayan city of Tikal in Guatemala, faces west towards the setting sun. Built to honour the end of each day, the Temple was seen by the Maya as a portal to the underworld.

moons, and seasons, and they extended the idea into death-and-rebirth cycles in their myths and legends. This meant that some dates were auspicious for particular events or ceremonies. When a particular cycle was complete they often erected monuments to mark the occasion. One of their great cycles began in 3113 BC and is scheduled to end on 12 December 2012, when they believed not that the end of the world would come, but – surprisingly – that there would be a reversal in Earth's magnetic field.

The Aztecs were a group of peoples who lived in central Mexico in the 14th, 15th, and 16th centuries, until they were conquered by Spanish invaders in 1521. The Aztecs invented calendar wheels, which were both a way of keeping time and also the essence of time itself – cyclical. Taking some of their ideas from the Toltec people and the Maya, they had even stronger beliefs in the cyclical nature of time. One cycle lasted 18 months, and for each of these months there had to be a ritual sacrifice to appease the gods. Many of their sacrifices were human – their hearts were cut out and held up to the Sun – and some historians estimate that thousands of victims were sacrificed every year. When they reconsecrated the Great Pyramid of Tenochtitlan in 1487 they claimed to have sacrificed 84,000 prisoners in four days.

BELOW: To sacrifice 84,000 in four days, Aztec priests would have to have worked around the clock, killing 875 victims per hour, which means about one every four seconds.

Great Spirit of the Hopi

The Hopi tribe – "the peaceful people" – live on a reservation in Arizona. In their history the cycles of the rock and the plant gave way to the cycle of animals, which is coming to an end and giving way in turn to the human cycle, during which we will gain access to great powers. The Great Spirit who came down at the beginning of the animal cycle said that we should share the Original Teachings, inscribed on stone tablets, so that we "can live and have peace on Earth, and a great civilization will come about". In each successive world, however, people became increasingly unruly and promiscuous, and only the most obedient were led to the next world, where they could begin again, while all the others were destroyed.

Eastern time

Similar ideas cropped up in the eastern world. For both ancient and modern Hindus in India, time is cyclical and consists of repeating ages. The process of creation is also cyclical, involving both evolution and "involution", when time turns in upon itself. The shortest cycle is the maha yoga, lasting 4,320,000 normal human years, and divided into four ages. A thousand maha yogas make one kalpa, which is one day in the life of Brahma, who in turn lives for a hundred years (or 3×10^{14} human years). We are at the moment in the first Brahma day of the 51st Brahma year; so there is some time to go before all worlds are completely dissolved – but Brahma, and everything else, comes in cycles, and will return. Life and death are illusions; time is responsible for old age and death, and when we can overcome time, we will become immortal.

Recurring time

Some ancient Greeks believed that the world had an infinite past; that there was no beginning of time. Others had ideas of time cycles. Pythagoras, with his curious community at Crotona in southern Italy around 550 BC, believed in reincarnation. One of the reasons he did not allow his followers to eat beans was apparently his fear that he might be reincarnated as a bean. Even today some people believe they have lived before. Not long ago I was lucky enough to go to a reincarnation party in California. Among the illustrious guests were Cleopatra, James Dean, and Thomas Jefferson; all tempted along by the elegant invitation to "Come as you were".

Do these notions of cycles and the wheel of time help us to understand what time is? I am not convinced that they come down firmly on either side of the old Plato–Aristotle disagreement. Perhaps we should be content with the trite definition of American science-fiction writer Ray Cummings, who worked with inventor Thomas Alva Edison, and said in 1922, "Time is what keeps everything from happening at once."

ABOVE: An ancient Hopi prophecy says that when the Blue Star Kachina (ie Sirius) appears in the heavens, Purification Day will be at hand, and the Fifth World will begin.

WHEEL OF TIME

Many Hindu ideas were adopted by Buddhists. In particular the Wheel of Time or Kalachakra Tantra (right) is a complex five-chapter text in Tibetan Tantric Buddhism that contains philosophies and meditation practices. The first chapter deals with the method for calculating the calendar, and the birth and death of universes.

The entire Kalachakra tradition is embedded with cycles of time, from breathing in and out to the orbits of the planets. The Dalai Lama has given Kalachakra initiations around the world, and has said the public of today's degenerate society need to be exposed to the following tantra: work towards enlightenment using the energies of your body, always remembering that everything is under the influence of time.

As far as we know, time began with the formation
of the Universe in the instant of the Big Bang,
13.7 billion years ago.

ALL TIME

ISAAC NEWTON might have argued, if he had known about the
Big Bang, that time already existed, independently of any matter,
but most people would probably agree that the Big Bang was the
beginning of time. Before the Big Bang there was nothing: no space,
no place, no time.

Aeons

For astronomers, an aeon is a billion years (1Ga, for
giga-annum); that is less than a tenth of the age of the universe
(13.7Ga), but a quarter of the age of Earth (4.5Ga), and a million
millennia. Stars the size of our Sun have a lifetime of around 10Ga,
while larger stars have shorter lives, and may explode as supernovae.
Smaller stars last longer, and may become white dwarfs, but few have
yet reached the end of their lives. Within the context of conformal cyclic
cosmology the distant future of one aeon becomes the Big Bang of the next.

In general use, an aeon is just a very long time, much longer than an era, which
generally refers to a slice of history, either of rocks or of humans, which are on
entirely different time scales.

DLE: Our Sun is about
llion years old. In
ther 5 billion years'
, it will swell into a red
t, after which it will form
ectacular planetary
ula like the Eskimo
ula pictured here.

IMAGINING ALL TIME

his diagram represents the entire
history of the Universe, and of time
tself. The Big Bang started as a
singularity – a point – and a moment
ater the incredibly hot mass of
particles ballooned outwards
n a period of rapid inflation.
Within a few minutes this mass
cooled down to around 1 billion
Kelvin, and the first atoms were
ormed. Stars followed, and the
irst galaxies appeared after about
a billion years (1Ga). The universe
s still expanding, apparently being

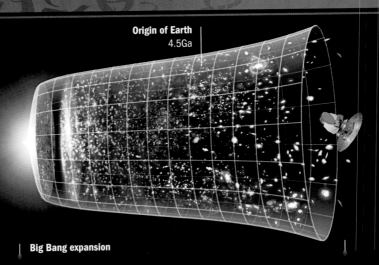

Origin of Earth
4.5Ga

| **Big Bang expansion**

The fossil record

The oldest rocks found on Earth so far are crystals of zircon from Western Australia, which are more than 4.4Ga old. Fossils (literally "things that have been dug up") are the remains of plants and animals, almost always species that have become extinct. Among the oldest fossils are stromatolites, which were once blue-green bacteria or algae and became compressed into layers of rock. The oldest undisputed fossils have been found in rocks that are 3.2Ga old, and the youngest, by definition, around 10,000 years old. Although the fossil record is incomplete it provides some evidence for the pattern of the evolution of life. Once scientists were able to estimate the age of the fossils, geologists were better able to date the rocks in which they were found.

ABOVE: The coelacanth is known as a "living fossil". Until recently, it was thought to have become extinct 64 million years ago, but this one was caught in the Indian Ocean in 1974. The species seems to have remained virtually unchanged for 350 million years.

The far future

What will life be like 100 billion years into the future? Let us imagine that on some Earth-like planet orbiting a star in our galaxy that doesn't yet exist, some intelligent creatures evolve. They discover mathematics, study science, and build telescopes, but they will never find out about the Big Bang. The Universe is expanding at an ever-increasing rate, and by the time these new scientists turn their telescopes on the heavens, the only stars in sight will be in the Milky Way, since all the other galaxies will by then be moving away so fast that their light can never get back. All evidence of the Big Bang, including the Cosmic Microwave Background Radiation, will have disappeared, so their studies will tell them, wrongly, that they live in a static one-galaxy Universe.

BELOW: The Cosmic Microwave Background Radiation (CMB), the afterglow of the Big Bang, has a temperature of 2.725K. In 100Ga it will have disappeared.

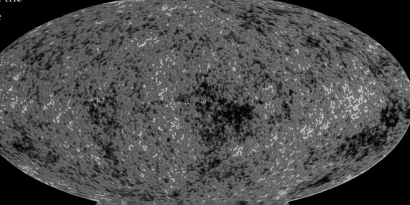

CHAPTER 2

NATURE'S CLOCKS

We tend to think of clocks as mechanical or electronic machines that count the passing seconds and measure the passage of hours for our benefit, but the world is full of natural timekeepers that control the behaviour and time-scale of everything from plants and animals to planets and stars. There are clocks in the heavens, clocks in the ground, clocks in the polar ice, and clocks in our bodies; all we have to do is look for them. Nature's clocks set the rhythms of life, and during the last 150 years scientists have teased out details of a vast variety of natural clocks, and have used them to unravel a few more secrets of the universe.

The Sun has always been the most conspicuous marker of time. Every morning it rises in the east and every evening it sets in the west. Even when there is thick cloud cover the arrival and departure of daylight is obvious, and the days are warmer than the nights.

THE SUN

ABOVE: At equinox on the equator the sun rises and sets in a vertical path, at right angles to the horizon.

BELOW: This time-lapse photograph taken in Alaska in summer shows the Sun dipping towards the horizon, but not below it.

MOST ANIMALS, INCLUDING HUMANS, use the Sun as a basic timekeeper, hunting, working, and otherwise living actively by day, and sleeping by night. A few creatures are nocturnal; badgers, burglars, and bats work by night, partly because this allows them to prey on the sleepy, partly because the world is quieter and cooler, and partly because there is less competition. In any case, each day and night marks one complete cycle, after which life begins again.

The rising and setting times of the Sun are not entirely simple. In the tropics the Sun tends to rise at 6am and set at 6pm all year round, and the day is 12 hours long, but nearer the poles the length of the day varies considerably. At the spring and autumn equinoxes, around 21 March and 21 September, day and night are each 12 hours long all over the world, but the further you go from the equator the greater the variation between winter and summer. Half way to the North Pole, at a latitude of around 45 degrees north, in the cities of Belgrade, Bordeaux, Harbin (China), Milan, Minneapolis, Montreal, Ottawa, Portland Oregon, and Venice, the length of the day (between sunrise and sunset) is nearly 16 hours in midsummer and less than 9 hours in midwinter.

At the North and South Poles the Sun never rises at all during the winter, but shines night and day during the summer months. Imagine that it is mid-September, and you are standing on the Antarctic snow near the South Pole at the end of a long dark winter. The sky has been getting gradually lighter for

A galactic year is the time it takes the Sun to o r b i t the Milky Way Galaxy.

weeks, and suddenly you catch a first glimpse of the Sun, shining through a distant valley. After a few minutes it disappears again, but the following day it reappears in the same place and stays in sight for a little longer. Within a week or two the Sun is describing low arcs across the horizon, longer and higher every day, until by December it is making complete circles in the sky. It never rises more than 24 degrees above the horizon – just over a quarter of the way up the sky – but it stays up there all day and all night. In fact "day" and "night" lose their meaning at the poles, as do "east" and "west"; from the South Pole every direction is north.

Wherever you are on Earth, however, the Sun appears to make one complete circuit in exactly 24 hours. In fact the Sun is not whirling round us; instead Earth is spinning on its axis, and the decision to call the time it takes for one revolution 24 hours was taken by many ancient peoples, including the Chinese, the Egyptians, the Indians, and the Sumerians. The 24-hour day–night cycle is now common throughout the world; meanwhile the day begins when the first flash of sunlight appears above the horizon, and ends when the last flash disappears.

That's about 230 million Earth years

THE SEASONS

Seasons result from the annual revolution of the Earth around the Sun, and the tilt of the Earth's axis relative to the plane of its orbit. For example, North America experiences summer between March and September because the northern hemisphere is tilted towards the Sun, and therefore receives more light and heat than it does during the winter, when it is tilted away.

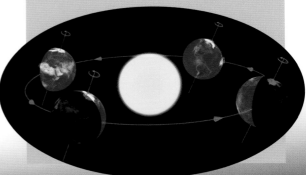

Early people, unencumbered by electric light and television, had plenty of time to look at the night sky and notice how the patterns varied from night to night and from month to month.

THE MOON

ABOVE: Although the Sun illuminates different portions of the Moon each night of the month, we always see the same "face" of the Moon, which does not change in relation to Earth.

THE MOST OBVIOUS OBJECT in the night sky is the Moon, which is sometimes a circular disc (full Moon), sometimes a semicircular disc (half Moon), sometimes a thin crescent, and sometimes does not appear at all, when it is called the new Moon, or the dark Moon. Some people reserve the name "new Moon" for the first visible crescent after the dark Moon.

The interval between successive full Moons is about 29 days, 12 hours, and 44 minutes (29.5306 days). This is not an integral number of days, which worried philosopher Nicolas Oresme (see p124), but because the Moon is so obvious in the sky, the idea of a month arose as the interval between full moons, and that remains roughly true; the calendar months average 30.4 days. This means that the full Moon comes about one day earlier each month; if there is a full Moon on 10 March there will be full Moons on or near 9 April and 8 May. Meanwhile if you

NEAP AND SPRING TIDES

Every 14 days, the Sun and Moon align with each other, and their combined gravity produces large "spring" tides. In between come the smaller "neap" tides, when the Moon and Sun are at right angles to each other in relation to Earth. Neap tides may be less than half as large as spring tides.

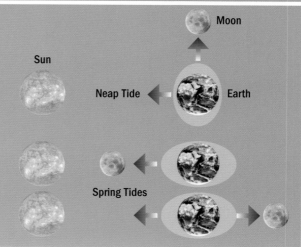

Moon

Sun

Neap Tide

Earth

Spring Tides

wait for the Moon to appear you will see that each night it rises a little later than the night before, as though it has slipped backward in the sky. The daily slippage varies between 15 and 90 minutes, but averages about 35 minutes.

Blue Moon

The expression "once in a blue Moon" means very rarely, or almost never. The Moon may actually appear to be coloured blue when atmospheric pollution from volcanoes or fires scatters red and orange light away. The expression, however, generally refers to the second full Moon in one calendar month. If there is a full Moon on, say, 1 May, then the next will happen 29.53 days later, which will probably be on 31 May. That second full Moon is called a blue Moon. Using this definition, a blue Moon happens about twice every five years.

Tide and time

The Moon is fairly close to Earth – about 380,000km (240,000 miles) away, although it varies slightly, because the Moon's orbit is not a circle but an ellipse. The Moon is also massive, and the effect of its large mass at such a close distance is a significant gravitational pull on Earth. This means that you weigh slightly less when the Moon is overhead than when it is not, although sadly the effect is too small to register on your bathroom scales.

BELOW: Because the Moon takes slightly longer than 24 hours to make a complete orbit, the interval between successive high tides is about 12 hours, 25 minutes. The tidal clock is of immense importance to sailors, lifeboat crews, and people who fish near the coasts.

The Moon's gravitational pull, however, is strong enough to cause the tides in Earth's oceans. As it passes overhead it pulls the water up into a slight hump, which follows around Earth just behind the Moon, and is seen by us as high tide. At the same time, Earth is pulled by the Moon, and leaves behind a hump of water on the side away from the Moon, where the Moon's pull is less strong. This forms a second high tide; so as Earth spins there are (in most coastal places) two high tides every day.

The Sun is much more massive than the Moon, but also much further away, and its gravitational pull on the oceans is smaller. However, when Moon and Sun are in a line with Earth, at full Moon and new Moon, the Sun's pull is added to the Moon's pull, and we experience more extreme tides, called spring tides (see opposite).

Following the regular movements of the Sun and the
Moon is relatively simple, but eclipses are much
rarer and more spectacular events.

ECLIPSES

WHEN THE FULL MOON PASSES through the shadow
of Earth it becomes gradually dusky, and a bite seems to have
been taken out of the disc. This bite grows larger, until the whole
Moon may become "swallowed", showing only a faint reddish disc
for some hours before it reappears in shining white. The red colour is
the result of scattering of light by Earth's atmosphere. When the Moon is
in Earth's shadow, no direct sunlight falls on it, but Earth's atmosphere acts as a fuzzy
halo, and lights the Moon. However, the molecules in Earth's atmosphere scatter
blue light, which is why the sky looks blue to us, and allows red light through. That
is why the Sun looks red at sunrise and sunset, when its rays have come through a
thick layer of atmosphere and much of the blue has been scattered away. And for
the same reason the eclipsed Moon is lit by light from which the blue has been
scattered away, and so it looks red.

ABOVE: The Moon is
swallowed up in Earth's
shadow. Light filtered
through the haze of Earth's
atmosphere gives the dark
part a reddish glow.

Solar eclipse

The Sun is eclipsed when the new Moon passes between the Sun and Earth, and
hides the Sun from Earth-bound viewers. You will see a shadow galloping across
the landscape towards you, and then suddenly everything goes dark – completely
dark if the Sun is entirely hidden in a total eclipse. This eerie effect lasts only for
a few minutes, but the event is dramatic. Animals and birds make nightfall noises,
and birds go to roost. Then the sunlight comes rushing across the landscape, and
everything comes back to life.

Predicting eclipses of the Sun and the Moon is not easy, but because the
events are so dramatic there has always been pressure to do so. In 2300 BC two
Chinese astrologers were beheaded for failing to predict a solar eclipse accurately.
Babylonian clay tablets record a solar eclipse on 3 May 1375 BC, and others
on 31 July 1063 BC and 15 June 763 BC. This last one is mentioned in the
Old Testament of the Bible (Amos 8:9). Babylonian astronomers seem
to have discovered a 223-month interval between lunar eclipses.

Thales of Miletus allegedly predicted an eclipse of the Sun in 585 BC (28 May),
and so stopped a war between the Medes and the Lydians. As the Greek historian
Herodotus wrote, "… day was all of a sudden changed into night. This event had
been foretold by Thales, the Milesian, who forewarned the Ionians of it, fixing for it
the very year in which it took place. The Medes and Lydians, when they observed
the change, ceased fighting, and were alike anxious to have terms of peace agreed
on." Since he was able to predict only the year of the eclipse, it seems surprising

THALES OF MILETUS (c.624–c.547 BC)

Thales, one of the earliest Greek philosophers, was born in about 624 BC and credited with all sorts of improbable achievements. Plato was scathing about him, saying that he always wandered around with his head in the air, looking at the stars or the weather, and that was why he fell into a well. Aristotle, on the other hand, said that Thales had been taunted – "If you're so clever, why aren't you rich?" So Thales used his meteorological knowledge to predict a bumper olive harvest, and bought up all the local olive presses, which he proceeded to rent out in a highly profitable season.

that the warring armies should have been so impressed, but perhaps even that was an amazing feat. An alternative, cynical view is that he was the only philosopher in the vicinity, and when the eclipse happened everyone assumed he had known about it in advance. Thales seems to have noticed that recent eclipses had happened every 17 years, and predicted that there would be 10 more in the following 170 years – and he was right, by chance. He did not understand the orbits of the Sun and Moon, although he did realize they are cyclical.

So the Sun and the Moon provide basic clocks for days and months, and their rare eclipses provided challenges for astronomers, and no doubt caused some worry and even panic among ordinary people, but could the heavens tell us anything else about time? Looking at the sky gave the ancients no clue as to the direction of time. The Sun, Moon, and stars seem to glide across the sky, disappear and return in endless cycles, the patterns repeating after days, months, years, or generations. However, the stars themselves do hold the answer to one of the greatest questions of all – when did time begin? The brilliant work of American astronomer Edwin Hubble is described in chapter five.

BELOW: Fiery prominences can often be seen past the edge of the Moon as it passes in front of the Sun during a solar eclipse.

An eclipse shadow moves very quickly – more than 3000km/h (1865mph) at the equator, and 4000km/h (2,485mph) at the poles

Edwin Hubble's discovery of the expanding universe (see p242) demonstrated that time runs in one direction – forwards. For some people, however, this was not obviously the case for time on Earth.

BIRTH, REPRODUCTION, AND DEATH

ABOVE: A bean plant sprouts from a seed.

MANY PLANTS APPEAR IN THE SPRING, flower during the summer, wither and die, and then appear again when the spring comes round. Perhaps this is why in many cultures people came to believe that life is cyclical – that events happen again and again, and people are reborn. But in reality this plant cycle is just another manifestation of linear time.

Plant a bean, and with luck – and absence of hungry herbivores – the seed will germinate and produce a seedling with cotyledons (first buds) that poke above the surface. As the weather gets warmer the stem will grow and develop leaves, and eventually flowers and fruit – the pods of beans that can be food for animals, including people – or the seeds for the next generation. Finally the plant will wither and die. Its life is complete.

This process is not reversible. You cannot push the withered remains into the soil and hope it will regenerate to form the living plant. It's a one-way system. On Earth, as in the heavens, time runs only forwards. Bigger plants may last for longer periods – some trees live for centuries – but still they start as seeds, grow steadily to maturity, and then wither and eventually die. No going back.

The same applies to animals. They are born either as live babies or as eggs. They gradually develop and grow. After some hours (fruit flies), months (tadpoles), or years (humans), they become mature adults, and then they gradually grow old and die.

Human beings can live more than 100 years – Jeanne Calment of France lived for 122 years, from 1875 to 1997 – but this is rare; most of us will die in our

MAXIMUM NORMAL AGES OF ANIMALS IN YEARS	
Mice	3
Chickens	10
Cats	21
Dogs	12–20*
Horses	40
Goldfish	49
Elephants	70
Giant tortoises	150
Whales	200

*depending on size

RIGHT: The giant clam can easily live for 100 years or more. The Icelandic Cyprine, another type of mollusc, is known to live for more than 400 years.

seventies or eighties. The human body seems to have a number of built-in ageing mechanisms, and various body systems – sight, hearing, digestion, brain – begin to work less well after about 60 years.

Modern medicine has enormously extended the average life-span, mainly by preventing death in infancy and from simple infections, but despite many optimists who think people may become immortal, the reality is that humans will never live much more than 100 years.

If the average human lives to be about **78 years** of age: that's **28,470 days,** or **683,280 hours,** or **40,996,800 minutes,** or **2,459,808,000 seconds**

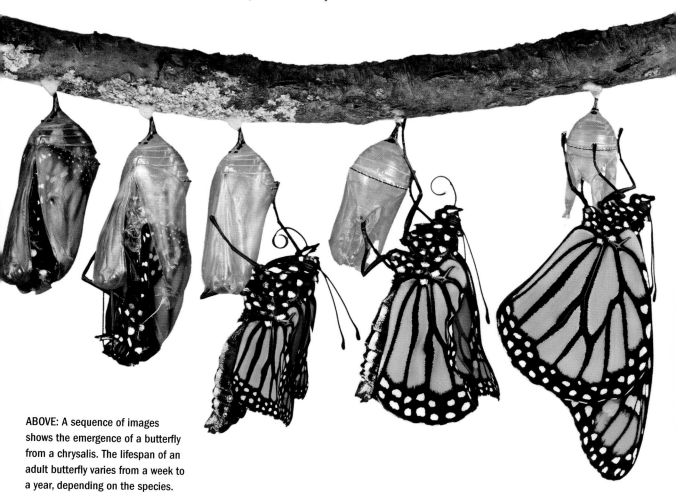

ABOVE: A sequence of images shows the emergence of a butterfly from a chrysalis. The lifespan of an adult butterfly varies from a week to a year, depending on the species.

For all the living things on Earth, nature's clocks tick away individual lifespans, yet hidden within the life cycles of all of them lies the process of evolution, which takes a much longer stretch of time than any individual life.

NATURAL SELECTION AND EVOLUTION

RECENTLY I CYCLED INTO A FOREST in central France. The grand, mature, deciduous trees towered overhead, with a canopy of beech, oak, ash, and other leaves, all different colours in the sunlight. On the ground, a carpet of dead leaves, and thousands of tiny seedlings, perhaps 10cm (4in) apart. As far as I could see, all these seedlings were of oak; no ash, no beech.

What was going on? Surely no one could have planted all those oak seedlings, or even uprooted all the ash and beech, even in the patch where I was standing. The result had to be natural. I would expect all the trees to produce their own seeds and scatter them liberally on the forest floor. Why, then, were the oak seedlings winning – and not only winning but flattening all the opposition?

One clue lay in the spacing: the mature trees were about 5–10m (15–30ft) apart. The seedlings were at least 50 times closer together. They could not all grow into mature trees: they would be too crowded, with the result that there would not be enough water or light to support them. It follows that even if all the mature trees were to die suddenly – or be cut down – barely one per cent of the seedlings would develop. In other words, 99 per cent of them would be doomed. Furthermore I saw no saplings; there were thousands of seedlings a few centimetres high, but nothing anywhere near a metre. Presumably all these seedlings are just one or possibly two years old, and all the older ones have already died, probably from lack of light and water, all of which is taken by the mature trees.

LEFT: The French forest has a lavish canopy of beech, oak, ash, and other trees, all competing for water and light.

I don't know what happens to the other species, but it is possible that the ash and beech seedlings never develop. The seeds may never even germinate as a result of lack of water. In that environment, oak is fitter, but still not fit enough to survive for long.

This is natural selection in action. Of all the millions of seeds produced in that forest every year, the vast majority fail to develop, for one reason or another. Similarly, in a river, a pair of fish may produce many thousands of fertilized eggs, but only two need mature in order to maintain the population of that species. More than 99.9 per cent of those eggs will fail to develop.

Which ones will survive? The fittest; the ones that are most suited to the environment. This may mean the toughest oak seedlings, or those in a small clearing, furthest from mature trees, where there is more available water and light. For the fish, the fittest may be the ones that prefer to stay under a rock, away from predators that are looking for breakfast, or those that develop into the fastest swimmers, or the greediest – the best at finding food. For each species, and for every environment, the needs of fitness will vary, but the same principle applies to all.

ABOVE: From tens of thousands of fertilized eggs, only a few fish survive to maturity.

Darwin's realization

Charles Darwin was the first to explain clearly the process of natural selection. Having read the famous essay on population by English demographer Thomas Malthus (see box), Darwin realized that all living things are ultimately short of food, and that most of them will die, if not from starvation then by becoming someone else's lunch. He pointed out, however, that within any family there is considerable variation. Siblings will always differ slightly from one another (unless they are identical twins), because the genes from their parents will have combined in slightly different ways. This applies not only to human children, but also to infant mice, fish, and oak trees. As a

> Which trees and animals will survive? The **fittest:** the ones most suited to the environment

---

THOMAS MALTHUS: *AN ESSAY ON THE PRINCIPLE OF POPULATION* (1798)

In his preface, Malthus says that he "first sat down with an intention of merely stating his thoughts to a friend... But as the subject opened upon him, some ideas occurred..." and the result was a book of some 400 pages. In chapter 1 he makes two postulates: "First, that food is necessary to the existence of man. Secondly, that the passion between the sexes is necessary, and will remain nearly in its present state." And he goes on to say that "the power of population is indefinitely greater than the power in the earth to produce subsistence for man. Population, when unchecked, increases in a geometrical ratio. Subsistence increases only in an arithmetical ratio. A slight acquaintance with numbers will shew the immensity of the first power in comparison of the second." He goes on to predict shortage of food leading to misery and vice. His book was almost entirely about humans, and led him to uncomfortable sociological conclusions, but it had immense influence. Malthus's fears have yet to be realized, however. Food technology has continued to sustain growing populations.

ABOVE: Darwin painted in 1881. He was rushed into publishing his book *On the Origin of Species by Means of Natural Selection* in 1859, when he realized that another biologist, Alfred Russel Wallace, was reaching the same conclusions.

RIGHT: An imagined progression of human evolution from an ape-like ancestor 3 million years ago.

result, all those oak seedlings have slightly different characteristics; some have bigger leaves; others are better at pulling water from the soil or at absorbing the energy from sunlight.

Let us suppose that water absorption is the most important feature for a seedling in that forest. Then the seedlings with the best water absorption genes will be likely to survive. More importantly, if they mature, they will pass on those genes to their children; so future generations will, in general, be better at absorbing water.

The evolutionary algorithm

The three critical processes are **variation** among the offspring – a range of characteristics among the siblings; **selection** between them – the environment causing water-absorption to be dominant; and **heredity** – passing on the useful genes to the next generation. American philosopher Dan Dennett described this as the "evolutionary algorithm". If you have variation, selection, and heredity, then you must get evolution. The species will gradually change. The process may take thousands of generations, but gradually one species may evolve into another, or into several others.

The best-known example must be the finches of the Galápagos Islands, heralded as one of Darwin's great discoveries during his voyage on HMS *Beagle*. In practice Darwin was so fed up with the voyage by the time they reached the Galápagos in 1835 that he shot a few birds on various islands, but put them all in one bag, and was induced to investigate properly only when the ornithologist John Gould pointed out what he had found – 13 different species of finch. There are about a dozen large islands in the Galápagos archipelago, and each island seemed to have its own type of finch, some with thin pointy beaks suitable for extracting seeds from fruit, others with thick parrot-like beaks that appeared to have been designed

for cracking nuts. Darwin wrote in 1845, "Seeing this gradation and diversity of structure in one small, intimately related group of birds, one might really fancy that from an original paucity of birds in this archipelago, one species had been taken and modified for different ends."

What seems to have happened is that some thousands of years ago a pair or more of finches were blown in a storm to the islands, which are 960km (600 miles) from Ecuador, the nearest mainland. Once installed, the finches bred, and spread from island to island, but the environments vary greatly. Some islands are almost deserts with volcanic vents and scrubby plants; others have a rich covering of green vegetation. In those varied environments the finches gradually adapted to their differing diets, with the fitter babies surviving in each generation on each island, until eventually, after thousands of repetitions of variation, selection, and heredity, they became different species. Nature's evolutionary clock ticks on long beyond a single lifespan to encompass the span of life itself.

ABOVE: Three of the finches collected by Darwin in the Galápagos Islands. All are descended from a common ancestor, even though one is much smaller than the others and their beaks vary greatly in shape.

If the human race lives **long** enough people will develop and change, in ways we cannot predict

The Very Reverend Dr William "Wild Bill" Buckland (1784–1856) was an English clergyman, a brilliant geologist, and a bit of a swashbuckler. His passion for natural history led him to one of the most startling discoveries of the age.

Dr Buckland's hyenas

IN THE 1840s BUCKLAND WOULD TAKE HIS STUDENTS on the new railway from Oxford, and, still wearing his academic gown, lecture about the geology of the passing countryside. He was something of an expert on dinosaurs and in the lecture theatre would strut like a bird to demonstrate how dinosaurs might have walked, which earned the disapproval of some of his more conservative colleagues. He boasted that he had eaten almost every member of the animal kingdom, and enjoyed them all except for the mole, which he said was disgusting.

The Kirkdale discovery

In 1821 in the tiny Yorkshire hamlet of Kirkdale, workmen went to fetch stones from the local quarry. There they found large numbers of bones, which they assumed had come from cattle that had fallen into the quarry and died. Buckland was excited by the bones from Kirkdale, for clearly they were not just those of cattle, and in December 1821 he travelled there to investigate. He found that in the side of the quarry was the mouth of a cave, which penetrated some 30m (100ft) into the limestone. He had to crawl through the entrance, but the cave opened up inside.

Buckland spent a week in the cave, and came away with a mass of information and a clutch of theories. He noted that all over the floor of the cave was a foot of mud, which he thought must have been deposited by the flood described in the Bible; for a clergyman this was an obvious assumption. Buried in the mud were thousands of bones – of rabbits and squirrels, but also bones of elephants, hippos, rhinos, giant deer, bison, and above all hyenas: he found the remains of more than 300 individual hyenas. As an additional puzzle, all the bones were broken – cracked through with jagged edges.

When the news came out, many theories were put forward to explain what might have happened. First among them was the notion that the flood was so terrifying that animals ran all the way from Africa, and hid in the cave to escape the rising water. The snag is that the entrance is so low that a human has to crawl through; even a very frightened elephant would have some difficulty getting in. And would

ABOVE: Dr Buckland dressed for glacier study and weighed down by the latest mineralogical kit.

rabbits really be scared enough to go into a cave with 300 hyenas? Besides, why were all the bones broken? Another theory was that the flood killed all the animals in Africa, smashed their bodies to bits, and swirled them all the way to northern England, where they fell into the cave through a hole in the roof. Problem: there is no hole in the roof, just several metres of solid limestone.

Buckland wrote a fascinating scientific paper about his findings in the cave, and came to a surprising conclusion. At one time, Yorkshire had been a tropical place, home to elephants and suchlike. Meanwhile the cave had been a den of hyenas; family after family had lived here and gone out hunting in the surrounding countryside. Hyenas are known to eat both freshly killed prey and also carrion. Therefore they might have dragged pieces of elephant back to the cave to finish off.

Deep time

The difficulty for Buckland lay in the timing. The number of hyena bones in the cave suggested that it had been their den for hundreds of generations, during all of which period Yorkshire must have been tropical. When could this have been? The Bible named only ten generations before the flood – not nearly long enough for the picture he had painted. James Hutton (see p71) had already suggested that deep time must have been needed for the sedimentation of rock strata; here was further direct evidence that Earth was much older than the 6000 years indicated by Biblical dates. In 1836 Buckland acknowledged that the Bible's account of the flood could not be confirmed by geology, and concluded that what he had assumed to be signs of the "Universal Deluge" was in fact evidence for the glaciation of an ice age.

There have been many creation myths about the origin of Earth, but one of the first attempts to calculate its age in a scientific way was carried out by Archbishop James Ussher of Armagh in Ireland in 1650.

THE AGE OF EARTH

ABOVE: James Ussher (1581–1656), born in Dublin, was Primate of All Ireland for 31 years. he was a prolific scholar, and also politically important in the 1640s, before and during the English Civil War.

AS A PRACTISING CHRISTIAN Ussher believed that everything in the Bible was true, and he used the family histories in the book of Genesis, and later the Age of Kings, to calculate the date of creation.

"First God made heaven and earth… God created man … male and female… When Adam had lived a hundred and thirty years, he became the father of a son… and named him Seth… When Seth had lived a hundred and five years, he became the father of Enosh…" The ages at which these chaps fathered babies seem rather improbable, but Ussher took it literally, and added up all the generations to work out how long had passed since the creation itself. His conclusion was that Earth had been created in the evening of Saturday 22 October, 4004 BC. Other scholars at around the same time calculated similar dates.

The trouble with these calculations was that they were based not on facts but on myths. The Bible was written by a variety of people over a period of probably hundreds of years. Even if they all thought they were recounting accurate history, the fact is that they were merely collecting old stories, handed down by world of mouth over many generations. Nevertheless many, at least in the Christian world, accepted these creation dates as fact. Other cultures had different ideas; the Vedic texts of the Hindus suggest that the universe is created, destroyed, and recreated in an endless cycle, each universe lasting for 4,320,000 years, while the creator Brahma has been around for some 158,700,000,000,000 years of human time.

Bishop James Ussher of Amargh calculated that Earth was created in the evening of

4004 BC

SATURDAY
October

22

Scientific dating

What has science had to say about the age of Earth, and indeed the universe? There has been a fascinating development of ideas and techniques over the last 2500 years. The philosophical poet Xenophanes of Colophon noted in about 500 BC

AL-BIRUNI (973–1048)

Born in what is now Uzbekistan in 973, Abu Rayhan Muhammad ibn Ahmad Biruni (or al-Biruni) was a Persian Muslim scholar, and one of the pioneers of experimental science. He was a friend and colleague of the philosopher Abu Ali al-Husayn ibn 'Abd Allah ibn Sina (Avicenna), with whom he had a prolonged scientific debate that was published as a book. He was the first Muslim to travel to India to study Indian philosophy and science, and in all wrote 146 books, most of them on some aspect of science. This diagram showing the phases of the Moon is from one of them.

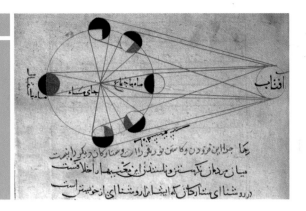

that the fossilized remains of sea creatures had been found at several places inland. He suggested that all the land had once been covered by sea, with the implication that this had happened long, long ago.

Fifteen hundred years later came the golden age of Islam, and similar ideas. The Persian scholar Al-Biruni described how sea shells and marine fossils were found on dry land, and reasoned it must once have been under the sea. Therefore, he said, Earth is constantly evolving, although it was too old to measure. Changes on the surface, he said, take a long period of time, the limits of which cannot be ascertained. At about the same time, the Persian scientist Avicenna suggested that layers of rock had been laid down one upon another.

Evidence from sea level

Some 700 years later, the French scientist Benoît de Maillet (1656–1738) picked up the same idea. He, too, had seen fossil shells in sedimentary rocks high above sea level, and assumed, like Xenophanes and al-Biruni, that all the land had once been covered by water. From places that had once been ports until the sea level fell he calculated the average rate of fall of water level to be about 0.75mm (0.03in) per year, and worked out how long it must have taken to fall from the peaks of the highest mountains. This gave him a time of two billion years since the land was entirely under water, and he estimated 2.4 billion years for the age of Earth.

BELOW: The fossilized remains of trilobites, small marine creatures from early in the history of life on Earth, unearthed in the western deserts of the United States.

Evidence from temperature

Georges-Louis Leclerc, Comte de Buffon (1707–88), was an energetic and scientifically minded author who published 35 volumes of *Histoire Naturelle* (Natural History). This was concerned mainly with plants and animals; a hundred years before Darwin's ideas he suggested that animals change with time. However, he also made an

If the age of Earth were compared with a *24-hour clock,* the first humans would appear only *40 seconds* before midnight

2 3 · 5 9 · 2 0

BELOW: By measuring the rate at which iron cools Buffon attempted to calculate the age of Earth. At the time he could not have reckoned on the effects of radioactive decay, which was discovered more than a century later.

estimate of the age of Earth, based on his own experiments. Buffon reckoned that if Earth had been a chunk of the Sun (see opposite), it must originally have been white hot, and must have taken a long time to cool down. In his lab in 1774 he heated balls of iron to white heat (perhaps 1500°C, or 2700°F) and then measured how long they took to cool to room temperature. From these experiments he was able to estimate how long Earth must have spent cooling down, and concluded that it was 74,832 years old.

Because Earth had been so hot, he reasoned, life must have begun in the far north, where the environment was cooler: probably in Siberia. Russian empress Catherine the Great was delighted, and rewarded him with furs, gold medals, and a diamond-studded snuffbox.

In private he reckoned that Earth must actually be much older than this, for aeons of time would be needed for chalk mountains to build up from the remains of sea creatures. He speculated on an age of about ten million years, but did not want to publish this estimate without evidence to back it up.

THE ORIGIN OF EARTH?

Newton had noted the improbable fact that all the planets revolve around the Sun in the same direction and almost in the same plane. Buffon calculated that the odds against this having happened by chance were nearly eight million to one. Therefore he reasoned (wrongly as it happened) that the planets must have been knocked out of the Sun by a comet, which would have made them all fly off in similar directions.

Evidence from rocks

James Hutton (1726–97) was a Scottish chemist and landowner, who became interested in geology during his experiments in farming. He studied rock formations around Scotland, and realized that much of the existing land had once been seabed, where layers of rock had been deposited and compressed. In many places these layers had been pushed upwards, so that they were now well above sea level, and often distorted, so that they were far from horizontal. He published his *Theory of the Earth* in 1788, and noting that there could have been many cycles of deposition, uplifting, and eroding, he concluded that Earth might be exceedingly old. As he wrote in the final sentence of his book, "The result, therefore, of our present enquiry is that we find no vestige of a beginning, no prospect of an end."

A fellow Scotsman and natural scientist, John Playfair, said, "the mind seemed to grow giddy by looking so far into the abyss of time." Hutton's ideas about geological time, also known as "deep time", were an important part of "plutonism" – his theory that rocks on Earth came from volcanoes, and had been modified by erosion and sedimentation; Pluto was the Roman god of the underworld, and therefore responsible for volcanoes. Hutton was also a founder of "uniformitarianism", the idea that the laws of nature have always been the same, and therefore the clues to the past lie in the present.

GEORGES CUVIER (1769–1832)

French naturalist and zoologist Georges Cuvier was brilliant at animal anatomy, and one of the first people to be able to put together skeletons of extinct animals to see what sort of animal they had been. Indeed he was the first person to show, in 1796, that some animals had become extinct. However, he believed that existing animals had always been the same, and that there was no such thing as evolution. He was also an early "catastrophist", suggesting that the changes to Earth, and extinctions of animals, had happened in a series of catastrophes, rather than by long slow processes, as proposed by James Hutton. One such catastrophe was the great flood, so vividly described in the Bible, and also mentioned in the Gilgamesh epic, a poem inscribed on clay tablets some 4000 years ago.

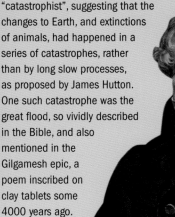

Back to the rocks

In 1841 Scottish geologist Charles Lyell visited Niagara Falls between New York State and Canada, and heard from a long-time resident of the area that the falls had moved upstream about 45m (150ft) since he had moved there 40 years earlier. In other words by eroding the rock underneath themselves, the falls retreat by about 1.1m (3.6ft) every year. He reckoned that this claim was probably exaggerated, and guessed at a real erosion rate of 30cm (12in) per year. That being so, he reasoned that the entire Niagara Gorge, some 11km (7 miles) long, could have been carved out by the river in 35,000 years. When he published this guess, it was accepted as an accurate measurement, although in fact he was quite wrong to assume the original claim had been an exaggeration. In fact the gorge is probably less than 10,000 years old, almost in line with Ussher's calculation of the age of Earth. Lyell may possibly have stuck to his estimate partly because he wanted to

BELOW: The rock beneath Niagara Falls is worn away by the flow of water at a rate of about a metre a year. This discovery allowed Charles Lyell to estimate the age of the gorge.

prove Ussher's estimate wrong. Nevertheless, that figure of 35,000 years was a real shock to those who still wanted to believe Ussher's 6000 years. The influence of Lyell's book was considerable. Captain Robert Fitzroy gave the first volume to Charles Darwin as they set off in HMS *Beagle* in 1831, and Darwin reported that he was seeing rock formations "through Lyell's eyes".

CHARLES LYELL (1797–1875)

Lyell, a Scottish geologist, was born beside a geological boundary fault separating farmland from the Grampian mountains. This sharp contrast in terrain may have been the first thing that sparked his interest in geology. He followed the ideas of Hutton, travelled to Europe and North America to see for himself, and wrote a ground-breaking book, *Principles of Geology* (1831–33). It provided support for the principles of Unitarianism; indeed the subtitle of Lyell's book was *An attempt to explain the former changes of the Earth's surface by reference to causes now in operation.*

As a student at Cambridge, Darwin had commented of one of his geology professors, Adam Sedgwick, "What a capital hand is Sedgewick for drawing large cheques upon the Bank of Time!" Later, however, in his famous book *On the Origin of Species by Means of Natural Selection* (1859), Darwin wrote that any reader who could not take in how much time was needed for biological evolution "may at once close this book". A year after Darwin's book was published, English geologist John Philips (1800–74) estimated the age of Earth at one hundred million years.

ESTIMATES OF THE AGE OF THE EARTH

	Age (thousands of years)	Date
Ussher	6	1650
de Maîllet	2,400,000	1738
Buffon	75 (or 10,000)	1774
Lyell	35	1831
Phillips	100,000	1860
Kelvin	20–40,000	1897
Current estimate	4,540,000	2000

Temperature again

William Thomson (1824–1907) was born in Belfast, Northern Ireland, educated at Cambridge, and at the age of 22 was elected Professor of Natural Philosophy at the University of Glasgow. He held the post for more than 50 years, and some of the experiments he began, on very slow diffusion, are still going on in the room that used to be part of his laboratory. In 1892 he was made a baron, and became known as Lord Kelvin (Kelvin is the name of the river that flows past his laboratory in Glasgow). He was a polymath, interested in everything.

He designed a machine for predicting the tides in all the harbours of Britain; he supervised both the design and the laying of the first successful transatlantic telegraph cable, as well as inventing the mirror galvanometer that made the use of the cable a practical proposition; he invented the binnacle for the successful housing of compasses in iron ships; he proposed an absolute scale of temperature, and in due course it was named after him: absolute zero is 0K (zero kelvin), and water boils at 373K.

Kelvin developed the idea of the second law of thermodynamics, which states that variations in temperature within a physical system will disappear over time. He used thermodynamic principles and the measured thermal conductivity of rocks to estimate how long Earth must have taken to cool from a molten state – a more scientific version of Buffon's idea of a century earlier.

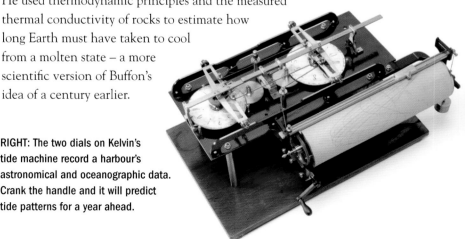

RIGHT: The two dials on Kelvin's tide machine record a harbour's astronomical and oceanographic data. Crank the handle and it will predict tide patterns for a year ahead.

His estimate of the age of Earth was somewhere between 20 and 400 million years. Later, in 1897, he revised this down to 20–40 million years, which was just about the shortest age possible from the geological record. He thought it could not be much older than this, in spite of the evolutionary evidence of deep time advanced by Charles Darwin, because he did not believe that the Sun could have been shining any longer than that.

Radioactivity had only just been discovered in 1897, and Kelvin did not know that the Sun is powered by nuclear reactions, nor that radioactive decay of minerals within Earth's crust provides heat that greatly slows the rate of cooling. Nevertheless, Kelvin's estimate of Earth's age was respected for decades.

ABOVE: A volcano spews molten rock, heated in the mantle, partly by radioactive decay. Realization of this changed the perception of the age of Earth.

KELVIN GAFFES

Kelvin was fond of making profound statements, some of which were gloriously wrong. He said that X-rays were a hoax, until he saw the evidence for himself. He said in a 1902 newspaper interview, "No balloon and no aeroplane will ever be practically successful." And he is widely reported to have said in 1900, "There is nothing new to be discovered in physics now. All that remains is more and more precise measurement." He did not foresee that the new sciences of quantum mechanics and relativity would appear on the scene within five years.

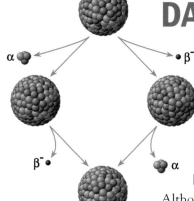

The oldest rocks yet discovered on Earth are crystals of zircon from Western Australia, which are more than 4.4 billion years old. Surprisingly, however, some of the best evidence of the age of Earth comes from the radioactivity of extraterrestrial lumps of rock.

DATING TECHNIQUES

MOST OF THE SOLAR SYSTEM seems to have formed at about the same time, and some of the rocks that did not get buried in the planets are still around as comets and asteroids. A few of these fall to Earth as meteorites, and some are known to be 4.54 billion years old, which is now assumed to be the age of Earth.

ABOVE: An example of the radioactive decay chain of a heavy element as it falls apart naturally through nuclear fission.

Radiometric dating

Although "radiometric dating" sounds like a lonely-hearts radio programme it is in fact the primary method for dating not only Earth but many ancient objects. It depends on the rate at which radioactive materials decay, for radioactive atoms make reliable timers. They are among Nature's most precise clocks.

Radioactivity was discovered by accident in 1896. French scientist Antoine Henri Becquerel thought that crystals of a uranium salt would absorb energy from sunlight and then emit light later by phosphorescence. The idea seemed to work when he detected the emission by the image left on a photographic plate. However, the following day was cloudy; so he could not repeat the experiment. He left the crystals on a photographic plate in a drawer, and again they left an

BELOW: Henri Bequerel, who discovered radiation by accident. In 1903 he shared the Nobel Prize in Physics with Pierre and Marie Curie.

image. He realized that there must be an internal source of energy, which turned out to be the result of the breakdown of atoms of uranium. The term "radioactive" is in fact misleading as no radio waves are produced. Rather, radioactive decay yields what came to be called alpha, beta, and gamma rays.

Many heavy elements are unstable. The nucleus, which is the lump at the centre of each atom, becomes "top-heavy" if it has more than about 91 protons – all have the same positive charge and so repel one another. A typical example is an isotope (one particular type of atom) of uranium with 92 protons and 143 neutrons, making a total relative atomic mass of 235. Scientists write this as ^{235}U. This isotope comprises less than one per cent of naturally occurring uranium, but it is radioactive, which means that, sooner or later, an atom will fall apart spontaneously, in a process called nuclear fission, and eventually turn into lead.

Radiometric dating was invented by New Zealander Ernest Rutherford, later dubbed the "Father of nuclear physics", while he was Professor of physics at McGill University in Montreal, Canada. Rutherford named the emissions from radioactive elements alpha, beta, and gamma rays, and he noted that radioactive atoms can act as clocks.

No one knows exactly when a single atom will decay, but if you had a million atoms of ^{235}U, you could be confident that half of them would decay in 704 million years. This is called the "half-life" of ^{235}U. After a further 704 million years, half of the rest would have decayed, leaving only one quarter of the original amount. Meanwhile its big brother ^{238}U, another isotope of uranium, which has three extra neutrons and comprises 99 per cent of natural uranium, has a half-life of 4.46 billion years. So if you find a rock sample in which half the ^{235}U has decayed to make lead, it must be about 704 million years old; if half the ^{238}U has gone, it must be 4.46 billion hears old. This is the method scientists use to date those chunks of rock and meteorites.

ATOMIC NUMBERS AND ISOTOPES

An element is a chemical substance that cannot be broken down by chemical processes into a simpler substance. Each element has a characteristic atomic number, which is the number of protons in the nucleus of each of its atoms. Examples are hydrogen (1), carbon (6), nitrogen (7), oxygen (8), and iron (26). In addition to protons, atomic nuclei contain neutrons, and all elements have a mixture of atoms with various numbers of neutrons. Hydrogen atoms always have one proton, but may have 0, 1, or 2 neutrons; carbon atoms always have six protons, but may have 6, 7, or 8 neutrons. These atoms with different numbers of neutrons are different isotopes of hydrogen and carbon. Iron atoms may have any number of neutrons between 28 and 34, so there are seven isotopes of iron, of which four are radioactive.

BELOW: The construction of Ale's stones in Sweden was dated to AD 600 by the radiometric dating of carbon material found at the site. The rocks themselves are hundreds of millions of years older.

Radioactive atoms act as clocks, allowing the oldest RocKS in the solar system to reveal their age

DATING THE EARLIEST TOOLS

The first tools used by early humans were made of stone, and cannot be dated directly. However, the materials surrounding finds of ancient tools can provide good evidence. Stone axes and knives from Gona in Ethiopia have been dated precisely in this way. The layer of volcanic ash on top of them is 2.5 million years old, while the magnetic traces in the rock below show that it is 2.6 million years old.

A good example of radiometric dating is the dating of rocks of the mineral zircon from the Pfunze Belt in Zimbabwe. Zircon contains a small percentage of uranium salts, and by measuring the fractions of both ^{235}U and ^{238}U that had decayed to form lead, scientists were able to establish the age of the rocks as 2.6 billion years.

A closely related system is provided by potassium and argon. The element potassium occurs in many minerals, and one particular isotope of potassium, ^{40}K, is radioactive. It decays with a half life of just over a billion years to form small amounts of the gas argon (Ar), which escapes from liquid rock but is trapped when the rock solidifies. The relative amounts of ^{40}K and argon in a rock can be measured and used to calculate the age of the solid rock. This works particularly well for volcanic lava flows. While the lava is liquid the argon escapes, but when it solidifies the clock is set to zero, and from then on the argon is trapped.

BELOW: The Oldupai Gorge in Tanzania, where archeological finds were dated by the radiometric evidence of lava flows.

The age of archeological remains in the Oldupai Gorge in East Africa was estimated by K-Ar dating of the lava deposits above and below them.

Living things are made of organic compounds, with chemistry based on the element carbon, and one isotope, carbon-14 or ^{14}C, is radioactive, with a half-life of 5730 years. This ^{14}C decays in dead material, so the amount of radioactive ^{14}C that remains can be used to calculate when an animal or plant died. The half-life of ^{14}C is much shorter than those of the uranium atoms above, so dates can be measured only up to about 60,000 years ago. Nevertheless this technique, developed by Willard Libby in Chicago in 1949, has been a powerful tool for dating all sorts of human artefacts, from an ancient Egyptian barge whose age was already known, to the Shroud of Turin, which was shown to be only about 700 years old.

Radiometric dating revealed the Turin Shroud, long considered a holy relic, to be a medieval fake dating from the 14th century

Dating by light

One pretty way of obtaining dates for ancient artefacts is by thermoluminescence (TL). When a piece of pottery lies around for hundreds of years it is bombarded by cosmic rays and radiation from radioactive minerals. This bombardment causes some electrons inside the material to get trapped at higher energy levels, rather like oversized basketballs that are too big to drop through the hoop and get trapped on top. When the pot is heated in the laboratory, the trapped electrons drop back to their proper energy levels, and as they do so they emit a minute amount of light. The amount light emitted gives a measure of the age of the object.

ABOVE: Some crystalline materials absorb energy from radiation, and then glow when heated in the lab. This thermoluminescence can be used to date the crystals.

Dating by handedness

When living things die, all sorts of chemical reactions begin in the tissue. One of the slower reactions is the process leading to even-handedness in the proteins. All living things contain proteins made of long chains of amino acids, which can be built in two ways that are mirror images of each other, like right and left hands. All but one of the 22 normal amino acids are left-handed. (The remaining one, glycine, is ambidextrous – neither left- nor right-handed.) When the organism dies, the amino acids slowly switch handedness until, after many thousands of years (depending on the temperature and other factors), there are equal numbers of right- and left-handed amino acids. This is called a racemic mixture, and the process is called racemization.

A VARIETY OF TECHNIQUES

Archaeological samples may vary in age from a few hundred to a few million years old. No single technique can embrace this wide range. Radiometric dating is in principle precise. A sample of radioactive material stored in isolation could be dated accurately. Real samples, however, have usually been exposed to wind and water, and some of the critical ingredients washed away. As a result most dating techniques are subject to error, and scientists have developed a range of methods, partly because some work better than others in particular conditions, and partly so that results can be cross-checked.

From the degree of racemization of the amino acids, scientists can estimate the age of the specimen. This process helped British scientist Simon Parfitt and his colleagues to estimate the age of artefacts found at Pakefield on the east coast of England in 2005. These suggested that early humans, perhaps *Homo Heidelbergensis*, were making flint tools there 700,000 years ago.

Dating with rings

A much simpler method of dating is to count tree rings. As trees grow their trunks get gradually thicker, and when they are cut down the growth rings are often obvious. Every year the tree grows during the summer and stops growing in the cold of winter, and the result is that a new ring is formed every year. The idea of using these rings as a serious dating method was first developed by American astronomer A E Douglass at the University of Arizona. He was studying sunspots, and reasoned that years with many sunspots would give rise to unusual tree rings; therefore he could use tree rings to investigate the history of sunspots.

BELOW: The age of a tree is given by the number of rings. Thicker rings show years of robust growth; thinner rings reveal meagre years.

OPPOSITE: Pueblo Bonito is the largest Great House in Chaco Canyon, New Mexico. Occupied for about 300 years from AD 828 by Ancient Pueblo people, or Anasazi, and built right underneath a leaning canyon wall, it is cut in half by a central wall, with a great kiva on either side. In all there were about 700 rooms on four or five floors. It may have been a village, or a city, or perhaps mainly a place for rituals and ceremonies.

The technique of dating using tree rings, known as dendrochronology (from the Greek *dendros*, a tree) has subsequently been used for all sorts of artefacts. Some trees live for many hundreds of years – bristlecone pines for several thousand – and examination of their rings yields patterns of growth. For example six years of energetic growth, producing wide rings, may be followed by two years of poor growth (narrow rings), one of good, and then three more of poor. If the same pattern can be found in other trees it provides a sound marker system. By comparing rings from many trees, scientists have been able to extend these markers back for thousands of years and build a reliable reference bank, which allows remarkably precise dating.

Tree rings revealed that the cultures of the Chaco Canyon, New Mexico, began building in about... AD 850

A timber from a building or a ship can be examined, for example, and from the pattern of the rings scientists may be able to fix an exact year in which the tree was still growing, which gives a maximum age for the building. For example if a portion of the pattern in a timber shows that the tree from which it came was growing in 1756, then that tree must have been cut down in 1756 or later, and the building built some time after the tree was cut down.

This has been especially useful in the dating of buildings used by American Indians. Around 1930, Douglass used dendrochronology to show that Pueblo Bonito, the largest Great House in Chaco Canyon, New Mexico, is about 1100 years old. The good news is that you do not have to saw right through the timber to count the rings. Instead scientists drill out a thin core from the outside to the centre, and count the rings in that core, without destroying the entire timber.

BELOW: Bristlecone pines are the oldest single organisms on Earth. Some have lived for more than 5000 years.

Dates from ice

In Antarctica, Greenland, and other places such as high mountain tops, the temperature never gets up to 0°C (32°F), the melting point of ice. Snow falls every year, and settles in layers, which build up year by year, and are gradually compressed by the pressure from above into a type of ice called "firn",

which looks like granulated sugar. As the pressure grows even higher the firn is compressed into solid ice. Within the ice are trapped tiny samples of gases from the atmosphere, dust, volcanic ash, and so on.

German scientists were the first to attempt serious study of the ice layers. In 1930, as part of the Alfred Wegener expedition to Eismitte (ice centre) in central Greenland, glaciologist Ernst Sorge studied the snow and firn layers in the sides of a pit 15m (50ft) deep that he dug by hand. His work showed that this could be a useful area of research.

Since then there have been many investigations all over the world. The normal method is to drill an ice core and examine it at leisure on the surface. The drill has a circular bit mounted on the end of a tube about 10cm (4in) in diameter – as wide as a drainpipe. When the drilling is complete the entire column of ice is brought to the surface, pushed out of the tube, and stored. In the 1960s, at Camp Century in Greenland, an ice core was recovered from a depth of 1390m (4560ft). The longest ice core recovered so far is 3270m (10,700ft), which provides an unbroken record of atmospheric conditions stretching back 740,000 years. This was the result of the European Project for Ice Coring in Antarctica (EPICA), a multinational European project with contributions from Belgium, Denmark, France, Germany, Italy, the Netherlands, Norway, Sweden, Switzerland, and the UK.

BELOW: Cores from the Greenland ice sheet are stored at a temperature of –20°C (–4°F). They will be examined for evidence of global warming.

WHAT CAN WE LEARN FROM ICE CORES?

Ice cores give information about particular years, or more often about groups of years. When the ice of a particular slice is analyzed, it can provide details of:

- local temperature at the time (from the released gases), and thus indications of glaciations and global warming
- concentrations of greenhouse gases including carbon dioxide and methane, showing how much of each was in the atmosphere at the time
- volcanic ash, and therefore volcanic eruptions
- organic ash, indicating forest fires
- sand and dust, after periods of desertification
- manmade chemicals, from organo-halogen compounds, first made in the 1880s, to chlorofluorocarbons (CFCs), made in large quantities in the second half of the 20th century, but now banned because of the damage they do to the ozone layer in the atmosphere
- atmospheric nuclear weapon tests, which generate compounds containing atoms of chlorine-36 (^{36}Cl) in much higher concentrations than usual.

Retrieval and examination of ice-cores has become a major source of information about the atmosphere in past eras, and the cores have proved to be one of nature's most useful clocks.

The upper parts of these ice cores can be dated by eye. The layers are clearly visible down to about 1800m (6000ft), and can be counted like tree rings. Below a certain level, however, the layers are so compressed that they become solid ice, and dating has to be indirect. One way of doing this is by looking for layers of volcanic ash in the ice. There have over the centuries been a number of well-documented eruptions that had world-wide effects on climate, and the ash from these shows up clearly in the ice, providing dates for particular depths.

BELOW: An ice core drilled from a depth of 1837m (6026ft) reveals a detailed atmospheric record in distinct light and dark layers.

Ice layers are a **TIMELINE** of Earth's climatic history, recording ancient volcanic eruptions as well as modern nuclear explosions

What can these scientific dating systems tell us about the beginning of life on Earth? The evidence is indirect.

WHEN DID LIFE BEGIN?

NO LIVING THINGS THAT WERE

ALIVE billions of years ago are still around now, but some of them have left traces. Fossils are the remains of living things, often bones that have been preserved in mud or turned to stone, and some of the best-known are ammonites. The oldest undisputed fossils have been found in rocks that are 3.2 billion years (Ga) old. That means there must have been life on Earth then.

There may have been fossils before 3.2Ga ago, but they would probably have been destroyed by the melting and squashing of the rocks. However, William Schopf of the University of California at Los Angeles suggested in 2002 that some stromatolites (layered rocks) contain fossilized bacteria 3.5Ga old, when Earth itself was only a billion years old.

Manfred Schlidowski and Stephen Mojzsis have argued that 3.85Ga-old samples of graphite found on Akilia Island in Greenland bear the hallmarks of organic origins. The evidence is subtle, and depends on a slightly low percentage of the isotope carbon-13.

Some scientists disagree, and suggest that the low ^{13}C content might have been caused by something other than living organisms, and also that the graphite was not sedimentary. The jury is still out on this case (see p89).

Most scientists agree that if life did begin on Earth (rather than in outer space) then it probably happened in water, since water is not only necessary for all life on Earth but is also capable of dissolving many vital ingredients and bringing them together in solution.

ABOVE: The calcified fossil of an ammonite shell. Ammonites were squid-like creatures that swarmed in vast numbers through the world's oceans 440 million years ago. They are excellent "index fossils", allowing geologists to date layers of rock.

MIDDLE: Stromatolites in Shark's Bay, Australia, are the fossilized clumps of ancient bacteria. They are among the oldest evidence for life on Earth.

HOW DID LIFE BEGIN?

No one knows for sure how life began. In 1871 Darwin speculated that in "some warm little pond, with all sorts of ammonia and phosphoric salts, light, heat, electricity, etc. present, that a protein compound was chemically formed..." In the 1920s it was suggested that life might have begun in the atmosphere, forming a "primordial soup" in the ocean. In the 1980s astronomers Fred Hoyle and Chandra Wickramsinghe saw what appeared to be evidence for organic molecules in space dust, and suggested that life arrived on Earth inside meteorites or comets. Recently German chemist Günter Wächtershäuser has suggested that life began beside a hydrothermal vent (right) in the ocean, where there is hot water, dissolved carbon dioxide, and plenty of minerals.

Geological evidence shows that oceans appeared within 150 million years of Earth's formation, so that sets an early limit to the beginning of life. The environment then would have been hazardous. Not only would Earth still have been hot, but there were frequent bombardments by asteroids, which could well have vaporized the newly formed oceans and created heavy cloud cover, blocking out sunlight. This suggests that life could scarcely have survived on the surface before about 4Ga ago, although if it started in the deep ocean, near a hydrothermal vent, it might have survived from 4.2Ga ago.

The first oceans
appeared 150 million years ago, and life... very soon after

So the consensus is that life began around four billion years ago, when Earth was half a billion years old. That early life was probably some single-celled slime, and nothing to write home about, not that there was anyone to write, nor anywhere called home.

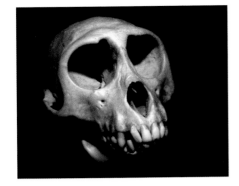

ABOVE: A reconstructed skull of *Australopithicus Afarensis,* one of the earliest hominids.

When did the first humans appear?

This depends on what counts as human. There is much controversy about human origins, but in the process of evolution it seems that human-like creatures of some sort diverged from their common ancestor with chimpanzees about 6 million years ago. *Homo habilis* appeared in south-east Asia, *Homo erectus* in Asia and *Homo sapiens neanderthalensis* in Europe. *Homo sapiens* in the form of modern humans seem to have appeared in Africa about 200,000 or 250,000 years ago, and lived for thousands of years alongside the Neanderthals, before the Neanderthals died out. Early humans learned to walk upright on two legs about 4 million years ago; they began to use simple stone tools at least 2.6 million years ago, and possibly as long as 3.9 million years ago; and they developed much bigger brains than other apes, possibly in

order to imitate one another and to communicate more fluently. By 50,000 years ago humans were enjoying language and music, and had begun to migrate out of Africa into the other continents.

BELOW: The *Cueva de las Manos* (cave of the hands) in Argentine Patagonia displays Stone Age human hands outlined by paint blown through a bone pipe. They were made some 10,000 years ago. They may have had a religious or a ritual purpose.

THE GREENLAND GRAPHITE

Graphite (right) is a form of pure carbon, which is an element found in all organic compounds: every living thing is based on carbon. In 100 atoms of carbon, on average 99 are carbon-12 (^{12}C) and one carbon-13 (^{13}C). Living things, however, concentrate the ^{12}C, so that the percentage of ^{13}C gets even lower. Graphite from Greenland was found to contain some carbon with significantly less than one per cent ^{13}C, which could be evidence of an organic history: it might have come from the remains of sea creatures fossilized long ago. This piece of graphite was in a rock that appeared to be sedimentary, with embedded chunks of zircon, which were dated radiometrically (see p76) to 3.85Ga. This implies that the graphite is at least this old, so the organic material is also at least this old.

We all have clocks built into our systems, including those that keep our hearts beating, our breathing regular, and our digestions ticking over. In other words, these biological clocks do not passively tell the time; they actively control biological systems within a time frame.

BIOLOGICAL CLOCKS

ANIMALS FOLLOW CONSISTENT PATTERNS of behaviour when it comes to eating, sleeping, mating, hibernating, and migrating. Bears and squirrels know when to hibernate and when to wake up; deer have clock telling them to mate in the autumn. Women have monthly clocks that control their menstrual cycles, and all humans have high-speed clocks in the cerebral cortex that tick 100 times a second.

Some of these cycles recur every day, but others come round only once a year, so there must be several types of biological clocks running at the same time. Some of these clocks are well understood, but others are not. The best-known cycle is the circadian rhythm, which encourages us to sleep at night and stay awake during the day. "Circadian" comes from the Latin *circa* meaning about, and *diem* meaning a day.

Circadian rhythms

Most living creatures, including plants and animals, are driven by circadian rhythms. We are so used to living from day to day that we scarcely notice them – except after a bad night's sleep – but scientists have been studying them for a long time. More than 200 years ago French astronomer Jean-Jacques d'Ortous de Mairan showed that mimosa plants open and close their leaves and spread their fronds in a particular pattern over 24 hours. They seem to do this in response to light, but he demonstrated that they continued to do so even when the plants were kept in complete darkness. Charles Darwin followed up this research, did more of his own, and suggested that plants go to "sleep" in this way in order to conserve energy. Today it is thought that they do so to conserve heat and water.

An interesting fact about human circadian rhythms is that when they are not triggered by light and darkness they tend to drift, and become a bit longer.

BELOW: A primrose opening its petals during the course of a morning, a process controlled by carbohydrates in the cell walls, and driven by daylight.

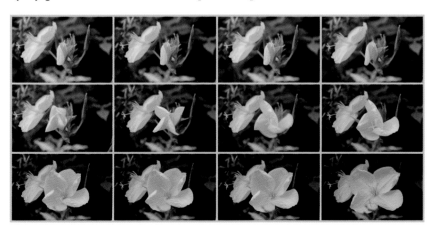

Your circadian rhythms are controlled by the suprachiasmatic nucleus in your brain, behind the **bridge** of your nose

CIRCADIAN RHYTHMS

In average human beings, the circadian body clock organizes a variety of functions: it produces the lowest body temperature at around 4am, switches on high blood pressure about 7am, triggers a probable bowel movement after 8am, brings the brain to high alert around 10am, and delivers maximum coordination and shortest reaction times in mid-afternoon, highest blood pressure around 6.30pm, highest temperature around 7pm, melatonin secretion from 9pm, and sleep before midnight.

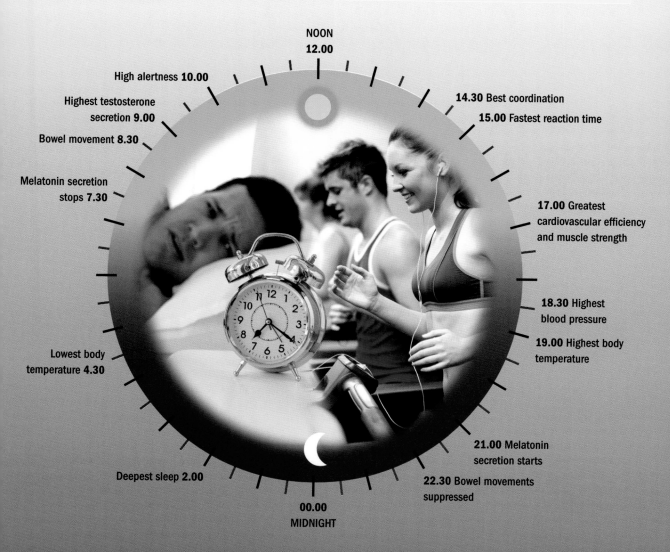

NOON
12.00

High alertness 10.00

Highest testosterone
secretion 9.00

Bowel movement 8.30

Melatonin secretion
stops 7.30

Lowest body
temperature 4.30

Deepest sleep 2.00

00.00
MIDNIGHT

14.30 Best coordination

15.00 Fastest reaction time

17.00 Greatest
cardiovascular efficiency
and muscle strength

18.30 Highest
blood pressure

19.00 Highest body
temperature

21.00 Melatonin
secretion starts

22.30 Bowel movements
suppressed

This implies that the genetic rhythms are not precisely in time with the Sun. However, they are brought back into step every day by the sunlight, or by some other synchronizer or time-giver, known by its German version *zeitgeber*. We know how the rhythms drift from a series of experiments in which volunteers lived in caves for extended periods (see p96).

ABOVE: Long-haul air travel can play havoc with your circadian rhythms. Many remedies have been proposed, but do they work?

Body clocks

Maintaining the rhythms is a complex business for the body, which produces a variety of hormones to keep everything in synch. One of the hormones involved is melatonin, which is normally secreted during the night, starting in the evening, two or three hours before you begin to feel sleepy. A study of teenagers in New York suggested that in the spring the production of melatonin in the evening was inhibited by the evening sunshine. This caused the teenagers to go to sleep 20 minutes later than usual and get 15 minutes less sleep, which made them tired and irritable in the morning. Melatonin is produced by the pineal gland in the brain, and it has been shown that when their pineal gland is removed, sparrows no longer show circadian rhythms.

Jet lag

Disturbing the circadian rhythm can have unpleasant consequences. One simple example is jet lag, caused by flying a long distance either east or west. Suppose you fly, for example, from Los Angeles to Paris; you get to your seat around 3pm and the flight takes 11 hours. When you arrive your body clock tells you the time is 2am. You want to be in deep sleep; indeed the cabin crew may have to wake you

THE MASTER CLOCK

The pineal gland and most circadian rhythms are controlled by the suprachiasmatic nucleus (SCN) in the hypothalamus. The SCN, which is actually a pair of glands, each about the size of a grain of rice and the shape of a pine cone, sits in the hypothalamus just above and behind where the optic nerves cross – that is about 3cm (1in) behind the bridge of your nose. Every neuron (nerve cell) in the SCN contains a biological clock, and all these cells run in synch to provide the SCN with its time. It sends out a continuous stream of nerve impulses, which is faster during active periods and slower during sleep. The SCN is the "master clock", but there are others, in the throat, lungs, liver, pancreas, spleen, thymus, and skin. Indeed it seems that most cells in your body have some way of modulating their activity in time with the day.

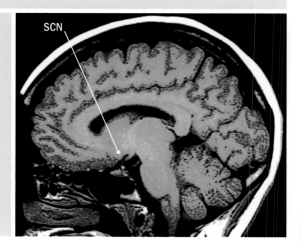

SCN

Many techniques have been suggested, and some may work. A simple one is to expose yourself to bright sunlight when you arrive. This is the most direct form of *zeitgeber* – a cue to tell you it is daytime. Another trick is to go to sleep as soon as you get on the aircraft – if you can. Refuse your airline meal and drink, and try to sleep. Then your body clock may be partly convinced that you have had a night's sleep, and it really is time to get up when you land.

Some people advocate a large meal, preferably with alcohol, before you board; this may help you go to sleep during the flight. Others advocate drinking during the flight, although this may make you feel even worse when you land, because alcohol causes dehydration; you would do better to drink lots of water. The same applies to coffee – it causes dehydration, and water is better.

Perhaps the best advice is simply to switch to the new time; reset your watch at the start of the flight, stay awake after you land, get out in the sunshine, eat meals at sensible times, and stay up till your watch tells you it is your normal bedtime. Then if you can sleep you should wake up with your body clock nearer the local time.

up before touchdown. You probably feel sleepy, groggy, disoriented, and dreadful. But Paris is nine hours ahead of LA; so in Paris it is 11am; the streets are buzzing, and for Parisians this is the time of high alertness and maximum activity. The contrast between how you feel and how they look makes you feel even worse, and your body clock may take two or three days to adjust to the local time – which means it is a mistake to go on such a trip for a weekend, since you will just about be recovering when you fly back and get into trouble all over again.

Jet lag makes you feel tired and irritable; it makes you restless and interferes with your sleep; and it often causes indigestion and feelings of nausea, because your digestive system is also controlled by your body clock, and wants to work at the "wrong" time. The nausea is caused by low blood-sugar levels, which should happen while you are asleep but now happen in the middle of the day.

The effects of jet lag depend on the length of the flight: the further you go the worse it feels, because the more your body clock is out of synch. Jet lag is supposed to be worse for older people, whose feeling for time is less elastic. I find jet lag is worse when I fly east than when I fly west, perhaps because flying west you just have a very long day; you do not lose a night's sleep. I once flew west from London to Vancouver. It was an eight-hour flight and an eight-hour time change, so we landed at the same time as we took off. This was disconcerting, although beautiful – we saw two sunsets during the flight – and it meant a long, long day, but I soon recovered.

One way to beat jet lag is to **sleep** on the flight – then your body will be fooled into thinking it's had a night's sleep when it arrives in a new time zone

RIGHT: Welders work a night shift to complete a new office building on schedule. They will attempt to sleep during the day.

Shift work

Most people work during the day, but about one in five work at night, driving trucks, firefighting, stacking supermarket shelves, sailing naval ships, making cars, and so on. The disruption to the body clock is made worse by the fact that the work times often cycle week by week:

Midnight to 8am > 8am to 4pm > 4pm to midnight

Working during the night is uncomfortable, because you have to sleep during the day, and therefore go against the suggestions of your body clock. Many shift workers say they never fully adjust – the inbuilt circadian rhythm is just too strong. Working rotating shifts is probably even worse, because you scarcely have time to get used to one pattern of working and trying to sleep at the wrong time before you have to change again.

One research programme showed that levels of melatonin, normally secreted at night, do not become synchronized with the new sleep patterns after a change of shift. Furthermore there is a build-up of fatty acids in

---

HOW AND WHY WE SLEEP

Most people go through two or three stages of increasingly deep sleep early in the night, and then progress to shallower, rapid-eye-movement (REM) sleep, during which they dream, while their muscles are paralyzed, so that they do not actually try flying out of the window and round the block.

As to why we need to sleep, the short answer is that no one knows. It may be to conserve energy, but in fact it does not conserve much: metabolism remains above 90 per cent of its daytime value. It may be to rest the brain and lay down memories. But nobody knows for sure. Sleep is necessary, however. If deprived of sleep, people get irritable, inefficient, and slow-witted.

the blood, which implies a risk of heart disease and diabetes. Because you feel tired and are less efficient, you are likely to lose concentration and make mistakes, which could be serious if you are in charge of a ship, or fighting a fire, on working in the emergency ward at the hospital. You will probably feel stressed, and you may well find that your sex life suffers.

Your health is also likely to be affected. Shift workers often report cluster headaches, which are said to feel like a red-hot poker inserted behind one eye. Women have said the pain is worse than that in childbirth. These headaches last from 15 minutes to three hours, and may recur several times a day.

Shift work goes against the body's  **sleep -wake cycle, resulting in poor concentration and health problems**

After some 15 years of shift working, people are four times as likely to suffer from ischemic heart disease. Women are more likely to develop breast cancer, because they produce less melatonin, which is a cancer suppressant. Social problems also arise, not merely because your social life is disrupted, but because of poor performance. The Exxon Valdez oil spill, the Chernobyl nuclear explosion, and the Challenger Shuttle disaster all happened because of mistakes made during night shifts. Perhaps they might not have happened in daytime.

BELOW: The Alaskan oil spill from the tanker *Exxon Valdez* in 1989 was a disaster in which the effects of shift work may have been a contributory factor.

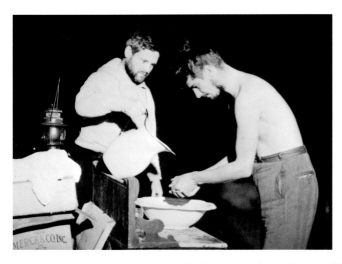

ABOVE: Kleitman (left) and Richardson taking a wash during the experiment.

A 28-hour "day"

In 1938 Nathaniel Kleitman, the "father of American sleep research" and his student Bruce Richardson spent 32 days 36m (119ft) underground in Mammoth Cave in Kentucky, trying to adapt to a 28-hour day. They had a regular schedule of eating, reading, exercise, and nine hours of sleep. The idea was to make a "week" of six days that would correspond to the normal week of seven, since 6 days of 28 hours make a total of 168 hours ($6 \times 28 = 168$), which is the same as seven days of 24 hours ($7 \times 24 = 168$). The point is that people living a 28-hour day would be awake for 19 hours and then sleep for 9; in 19 hours of wakefulness they could get more done than in a normal day of say 16 hours, which might be useful for students or military personnel. In their heroic experiment

Richardson succeeded but Kleitman failed, perhaps because at the age of 43 he was too old to be able to adapt easily, while Richardson was only 25.

Derk-Jan Dijk and Charles Czeisler tried further experiments with eight 20-something men over a month in 1994, giving them no time cues, but encouraging them to sleep for 9.33 hours every 28 hours. They showed that there is a sleep-wake cycle that is connected to but not enslaved by the body-temperature cycle and the melatonin-secretion cycle, and that waking up in particular seems to be controlled more by the sleep-wake cycle than by these others.

Aircrews and astronauts

During the early 1990s, scientists at the NASA-Ames Research Center made extensive studies of sleep patterns in military and civilian aircrews in order to learn about potential problems for astronauts. Under normal conditions on Earth, older people sleep less, and less deeply, during the night, and are sleepier during the day, than younger people. In other words they are less powerfully controlled by the sleep-wake cycle. They also tend to go to sleep earlier and wake earlier.

BELOW: Mammoth cave in Kentucky, where Kleitman and Richardson attempted to live a six-day week of 28 hours a day.

When normal circadian patterns were disrupted by long-haul flights, the research showed that older people (50–60) tended to lose more than three times as much sleep as younger people (20–30), and also that the daily temperature rhythm was weaker for older people with the minimum temperature occurring earlier. These results confirm the fact that older people are less able to adapt to disrupted rhythms and jet lag than the young.

Men and women have been going into space for 40 years, and many have stayed there for periods of many days or even years. All have been deprived of the normal circadian cues of light and darkness. The International Space Station, for example, orbits Earth every 90 minutes: its "days" are only 90 minutes long. Many astronauts have been wired and monitored extensively while in orbit, so they provide a considerable bank of information about free-running and artificial circadian rhythms.

Four male astronauts were monitored on space shuttle STS-78 orbiting Earth for 17 days. The mission was scheduled to minimize disruptions to circadian rhythms, so that the effects of space flight and microgravity could be studied. The result was that the circadian rhythms were similar to those on the ground, and roughly in time with the work-rest schedule. However, they did lose sleep, averaging only 6 hours 6 minutes of sleep in 24 hours.

Another study was carried out on five astronauts before, during, and after 10- or 16-day space missions. This time their work-rest schedule was imposed in periods of 23.5 hours. Light and darkness came variably on the flight deck, and the light was dim in other compartments.

EFFECT OF TEMPERATURE

The human circadian rhythm is not affected by temperature, but what about some of those high-speed clocks? We are remarkably good at perceiving where a sound is coming from (see p34) and the speed of a car coming up behind when seen in the rear-view mirror. For these systems we must have a high-speed reference clock, rather than just one that ticks every 24 hours.

Dr David Jones suggested that this high-speed clock must depend on chemistry, since the rates of chemical reactions are reliably constant, providing the body temperature does not change. When the temperature changes, however, chemical reactions go at different speeds – a 10°C (18°F) rise in temperature can double the reaction rate. Does that mean that by changing the body temperature you could change the rate of the clock?

The astronauts slept poorly, and for only 6 hours 30 minutes per day –
taking melatonin did not help – and their circadian rhythms of production
of cortisol in the urine became out of phase with the sleep-wake cycle.

One astronaut reported effects he noticed during a five-month flight
in the *Mir* space station. For three months he slept reasonably well and
stayed wide awake during the "day", while his body temperature fell as
he went to sleep and rose again when he woke. In the fourth month,
however, he began to lose track of each "day". He did not feel sleepy at
bed time, and his body temperature remained constant, showing
no sign of a cycle.

Astronauts may usually have regular work-rest cycles, but
every now and then they have to carry out some complex
task in the middle of their artificial night. For example another
spacecraft may need to dock at what is in effect 2am.
Research funded by the National Space Biomedical
Research Institute has used mathematical
models to calculate the best strategies
for individual astronauts faced with such
tasks: should they have an evening nap,
or should they drink lots of coffee to
stay awake? The results may
help shift workers on
the ground.

ABOVE: The "beds" on the
International Space Station
are like upright suitcases.
In zero gravity, however,
astronauts claim to find
them quite comfortable.

LEFT: Astronauts must be
ready to undertake complex
and delicate tasks regardless of
what hour it is for their body clock.

On the
International
Space Station, the
astronauts' circadian
rhythm remains roughly
the same as
on Earth

What would it be like to live deep underground? The first deliberate experiments by people to explore their circadian rhythms by living for a period of weeks in caves, without the *zeitgeber* of daylight, were carried out by Kleitman and Richardson in 1938 (see p96) and by Michel Siffre in 1962.

Troglodytes

IN AN INTERVIEW AFTER THE EXPERIMENT, Siffre, a French scientist, said, "You have to understand, I was a geologist by training. In 1961, we discovered an underground glacier in the Alps, about 70 kilometres from Nice. At first, my idea was to prepare a geological expedition, and to spend about 15 days underground studying the glacier, but a couple of months later, I said to myself, 'Well, 15 days is not enough. I shall see nothing.' So, I decided to stay two months. And then this idea came to me – this idea that became the idea of my life. I decided to live like an animal, without a watch, in the dark, without knowing the time." He was able to contact colleagues on the surface, but they did not contact him, and he had no clues as to the time of day.

Each "morning" – or rather when he woke up – he would phone the team, report his temperature (which became very low in the cool cave) and his pulse rate, and count from 1 to 120 in what he thought was one number per second, so that reaching 120 should have taken two minutes. In practice, after some days underground, he took five minutes to reach 120. His internal estimation of time seemed to have slowed down.

Without the cues from daylight, Siffre's circadian rhythm gradually drifted, and within a few days he was living days of 24.5 or 25 hours. Similar results were found with other volunteers – at Harvard University scientists found a natural rhythm ranging from 23 hours 55 minutes to 24 hours 27 minutes – and it seems that there is an inherent circadian rhythm that does not depend on cues from the world outside the body, and the "free-running" circadian rhythm (ie without light cues) for humans is rather longer

OTHER FREE-RUNNING CIRCADIAN RHYTHMS

- mice about 23.5 hours
- marmosets 23.6 hours
- hamsters 24.0 hours

The circadian rhythm is not affected by temperature; in other words there must be some compensating mechanism working to keep it roughly constant. Recent research shows that the control is genetic.

than 24 hours, but is brought back to 24 hours every day by the *zeitgeber* of light, or of social activity. The ideal light is blue, intense, and from above.

In 1972 Siffre spent 205 days in a cave in Texas. Knowing the results of the previous experiment, he probably attempted to compensate for the drift, but much the same happened again, except that once or twice he dropped into a 48-hour cycle, staying awake for 36 hours and sleeping for 12. This had previously been experienced by other troglodytes, but Siffre reported that he did not notice the difference between the 25-hour and the 48-hour "days".

One of the other findings from Siffre's experiments was that without the *zeitgeber*, the sleep-wake cycle and the body-temperature cycle get out of phase with one another. For most people, body temperature is usually highest shortly before bed time and lowest just before waking in the morning, but this shifted while Siffre was in isolation, so that his temperature dropped to its lowest point soon after he went to sleep.

Siffre's first cave experiment was essentially the beginning of the science of chronobiology, the study of time-cycles in living organisms and their relationship to the Sun and Moon.

RIGHT: Michel Siffre during the 1972 experiment in the Midnight Cave in Del Rio, Texas.

I live near a salmon river in Devon, England, and during the summer months, especially when there is a spell of dry weather, I can look down into the river and see the salmon waiting to swim upstream when the rain comes and the water is deep enough.

ANIMAL RHYTHMS

I HAVE SEEN A DOZEN SALMON in a space the size of a dining table, just lying motionless in the current as it rises over a slab of bedrock from the deep pool upstream. George, the old fisherman who has been fishing here for years, tells me they come up to spawn, and always the fishing is best around the time of the full Moon. I have seen two of the eight-pounders that he has caught, so he must know something about the salmon. Why do they come at the full Moon? Do they need the light to see their way, or is there some more primitive instinct?

BELOW: Salmon spend up to five years in the open ocean before returning to their natal stream to reproduce, often making amazing journeys for hundreds of miles against the current.

Riddle of the oysters

Oysters live in the space on the sea shore between the limits of high tide and low tide. When the tide is high they open their shells to feed, and as it goes down they close them again to avoid drying out. American marine biologist Frank Brown bought some oysters in his local fish market, and having satisfied himself that they

SOLUNAR FISHING

In 1926 John Alden Knight studied the effect of the Moon on fishing in Florida, and found that 180 out of 200 catches were made at the time of the new Moon. The Moon/Sun combination is thought to be important, and today there is a flourishing industry in "solunar" studies. You can buy solunar calendars, and expensive gadgets that will allegedly tell you the best times to catch fish.

showed interesting rhythms, he asked a colleague on the eastern seaboard of the United States for help. On 18 February 1954, oysters were dredged from a depth of 11m (35ft) near New Haven, Connecticut. Professor Brown had these oysters sent back to his laboratory so that he could study them further, and found that they still opened and closed in time with the tides of Long Island, even though there were no noticeable tides in the tanks in his laboratory. But what was more remarkable was that his lab was nowhere near Long Island, but a thousand miles to the west, at Northwestern University in Evanston, Illinois.

The oysters continued to open and close with the same rhythm for two weeks, but then suddenly changed. They remained synchronized with one another, but seemed to have slipped backward about an hour. He then realized they were opening in time with what would have been high tide in Evanston, if the ocean came that far inland.

Ocean tides are caused mainly by the pull of the Moon; as it passes overhead its gravitational force pulls the water upwards, causing a high tide shortly afterwards (see p57). Could the oysters be responding to the Moon? He kept them in the dark to eliminate clues from sunlight or moonlight, but still they maintained their rhythms. His conclusion was that they must be under the influence of the Moon's gravitational pull. Some deep instinct in these animals was triggered by minute changes in the gravitational field they experienced.

Oysters open and shut in response to the Moon

Life of a grunion

American environmental writer Rachel Carson recounted another amazing rhythm, in a fish called a grunion – "a small, shimmering fish about as long as a man's hand... Shortly after the full Moon of the months from March to August, the grunion appear in the surf on the beaches

ABOVE: Grunions at full Moon and spring high tide. They lay and fertilize their eggs in the sand, safe from the water until the next spring tide.

of California. The tide reaches flood stage, slackens, hesitates, and begins to ebb. Now on these waves of the ebbing tide the fish begin to come in. Their bodies shimmer in the light of the Moon as they are borne up the beach on the crest of a wave, they lie glittering on the wet sand for a perceptible moment of time, then fling themselves into the wash of the next wave and are carried back to sea." The reason for the grunions' brief trip to the beach is to lay and fertilize their eggs in the sand, where they will not be touched again by the sea until the next spring tide two weeks later. By then the larvae will be ready to hatch from their eggs, to be washed into the sea and swim to safety.

Microscopic commuters

Even some of the tiniest living things seem to respond to the pull of the tide – or more likely the Moon. A microscopic diatom called *Hantzchia virgata* is known as a "commuter diatom". It lives on sand and mudflats between high and low tide marks in North Carolina and Oregon, in the Netherlands and the UK. At low tide during the daytime it forms golden patches on the surface, while before high tide it burrows 2cm (1in) down, or deeper if there are fierce waves, and it stays down all night. When low tide gets near to sunset it switches, and does not come up until the next low tide in the morning. The burrowing could be a direct result of the water depth, but the little algae continue to move up and down in time with the tides even in a lab where there is no tide, and they go on keeping time with the Moon for several weeks. Either they have extremely accurate double clocks – one in synch with the Moon and one with the Sun – or they feel the pull of the Moon's gravity and are afraid of the dark.

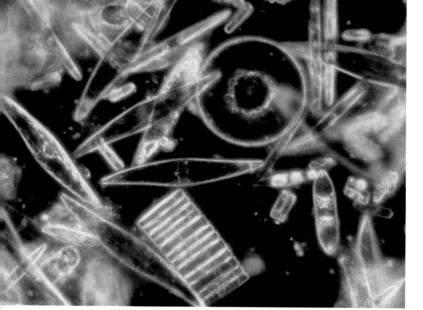

There is evidence that the Moon affects not only animals but also plants. Ernst Zürcher (see p106) has described how the diameter of trees increases and decreases in phase with the Moon, even when they are kept in complete darkness and at constant temperature. The change in thickness is only a fraction of a millimetre, but it is measurable and repeatable.

LEFT: These diatoms emerge from the sand only at low tide during daytime.

OPPOSITE: The Moon seems to exert a powerful effect on a wide variety of marine creatures, and also on plants.

In most of the last ten summers I have spent a week or so in the woods, making chairs, tables, and other things in green woodwork courses. During the process I have learned that when timber is wanted for furniture and similar uses, the trees are felled in the winter, while the Moon is waning; that is, between full Moon and new Moon (see p56–57).

PLANT RHYTHMS

APPARENTLY THE SAP RISES WHILE THE MOON IS WAXING, but falls while it is waning, so that the timber is less wet. This practice has been around for many generations: 2300 years ago Aristotle's successor Theophrastus of Eresos wrote in his *Causes of Plants* that there is an appropriate season for felling trees, and wood cut from those felled at the beginning of the waning Moon will be harder and less liable to rot. Similar practices seem to be common across the world today: in India, Sri Lanka, Brazil, Guyana, and the Middle East, as well as in Europe. There is plenty of superstition and folklore attached to the tradition, but scientific investigation backs it up.

Professor Ernst Zürcher, Chair of Wood Sciences in the Swiss Federal Institute of Technology, felled 30 spruce trees near Zurich during the winter of 1998–99, and found that immediately after felling there was no difference in density (weight per unit volume), but after drying those felled while the Moon was waning had 3–5 per cent higher density than those felled while the Moon was waxing. In other words the "waning wood" or "Moon wood" had less water to lose, and so was effectively drier. Two studies with the sapwood (the thin active layer under the bark) of European spruce gave Moon wood densities around 10 per cent higher after drying.

MOONLORE

Johannes Kepler (1571–1630) was a German astronomer who started his professional life as assistant to Danish scientist Tycho Brahe (famously the man with the silver nose) and went on calculate important astronomical tables and to work out the laws of planetary motion. In 1602 he wrote, "From experience is sure, that everything which is formed by humidity starts strongly growing with the waxing Moon, but diminishes with the waning Moon."

Roman author Pliny the Elder (AD 23–79) was a natural philosopher who wrote a large and comprehensive encyclopaedia of natural history. In it he said trees should be felled at the new Moon, and he advised farmers to pick fruit for market before the full Moon, as it weighed more, but to pick fruit for their own stores at the new Moon, since it would last longer.

Johannes Kepler

Compression strength, vital for building and woodwork projects, was found to be more than 10 per cent higher in Moon wood. Timber from waxing wood, however, turns out to be better for burning, producing more heat per unit of weight.

In 2010 Zürcher published the results of an enormous survey of 432 Norway spruce trees and 144 sweet chestnut trees, felled at 10am on Mondays and Thursdays between 6 October 2003 and 18 March 2004. The results are complex, and are analyzed for each part of Moon cycle, but basically they show that there are systematic variations in the properties of the wood that occur in time with the lunar cycles.

Tonewood, a Swiss company that grows timber for musical instrument makers, says that lumber was for centuries cut exclusively in the winter and then only at certain times in accordance with the phases of the Moon. According to them, "Years of experience show us very good results in working with Moon wood. We fell the trees only from late October until January, during the last days before the new Moon. The result is very light and stiff resonance tops; that's what our customers (instrument builders from all over the world) need."

And it is not just green woodwork and musical instruments. Trees have for generations been felled at the "right" time in the Moon's cycle for construction work, shingles, wine barrels, cheese boxes, longbows, ploughs, and even firewood.

ABOVE: Spruce trees in Scandinavia ready for felling during a waning Moon.

Lunar planting

Folklore says that transplants and grafts take best when the Moon is waxing and the sap is rising. The simplest rule is to plant crops that produce above ground during the waxing Moon, and those that produce below the ground while the Moon is waning. More specifically, plant spinach, brassicas, celery, and cress in the first week after the new Moon, and before the full Moon (ideally two days before) tomatoes, cucumbers, peppers, squash, peas, and especially beans. After the full Moon plant root vegetables – potatoes, parsnips, carrot – and onions, berries, bulbs, biennials, and perennials, but plant nothing in the week before the new Moon. Mow your lawn while the Moon is waxing, but prune trees and bushes (as with felling trees) while it is waning. (These suggestions follow the "synodic" cycle of waxing and waning, but there are more intricate methods, based on sidereal and biodynamic cycles.) As with fishing (see p102) you can buy Moon gardening calendar wall charts, Lunar gardening booklets, monthly digital downloads, and software to help you plan your gardening.

Professor Zürcher decided to investigate the effect of the Moon on planting, and carried out experiments on four different types of tree in Rwanda, where there is little variation in day length throughout the year, which eliminates some possible side effects. He planted 200 seeds for each tree, each in a separate compartment in a wooden crate, and measured how long the seedlings took to emerge, the rate of germination, and the height after four months. The results were impressive. Seeds sown two days before the full Moon (said to be the best time) germinated after an average of 47.5 days, while those sown just before the new Moon took 58.5 days – 23 per cent longer. (Radishes have also been shown to germinate faster when sown just before full Moon.)

ABOVE: The makers of musical instruments prefer to use "Moon wood" – from trees felled just before the new Moon.

Beans, peas, and tomatoes are said to **grow** best if planted in the second week after the new Moon

LEFT: Some serious gardeners choose their planting time according not only to the phase of the Moon, but also according to whether its path is rising or falling in the sky.

After four months, the average heights of the full Moon seedlings were consistently greater than those of the new Moon ones. In some cases they were twice as high. Similar results were obtained by Lilly Kolisko in the 1920s and 30s with cereals, vegetables, herbs, and flowers.

The variety of Nature's clocks is remarkable, and no doubt there are some that have not even been noticed yet. The subtle effects of the Moon rhythms on growing plants give some indication of just how many secrets remain for scientists to uncover.

PLANTING BY THE MOON

Phase	Plant
First week after new Moon	broccoli, cabbage, celery, cress, spinach
Second week	beans, cucumbers, peas, peppers, squash, tomatoes
First week after full Moon	berries, biennials, bulbs, carrots, onions, parsnips, perennials, potatoes
Second week	nothing

BELOW: Day by day, bean seedlings thrust upwards, and apparently they thrust faster if planted just before the full Moon.

Years are loosely defined as the interval between recurring seasons. Many ancient peoples built monuments to mark the shortest day of the year, and so were able to measure the year precisely.

YEARS

ONCE NICOLAUS COPERNICUS had persuaded us that the Earth orbits around the Sun, the year became the time taken for one revolution. Increasing precision of astronomy and scientific measurements made this too loose a measurement.

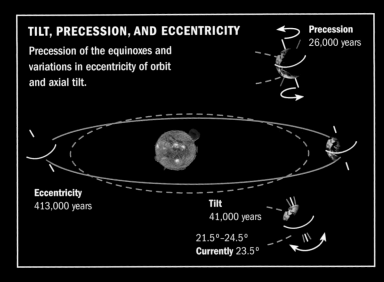

TILT, PRECESSION, AND ECCENTRICITY

Precession of the equinoxes and variations in eccentricity of orbit and axial tilt.

Precession
26,000 years

Eccentricity
413,000 years

Tilt
41,000 years

21.5°–24.5°
Currently 23.5°

Measuring a year

The "tropical" year, or solar year, is the time taken for the Sun to reach the same position in the seasonal cycle. This is the year that matches the Sun and our seasons. Because the Earth wobbles this is not exactly constant, but in 2000 its average was 365d 5hrs 48mins 45s, or 365.24219 days. The "sidereal" year (measured against the stars) is 20mins 24.5s longer than the tropical year (365d 6hrs 9mins 10s or 365.25636 days), mainly because of the precession of the equinoxes.

The Milankovitch cycles

While he was interned during World War I, Serbian mathematician Milutin Milankovitch calculated the effects on the climate of variations in Earth's orbit. The "precession of the equinoxes" is a rotation of the Earth's axis, like a wobbling

---

IMAGINING A YEAR

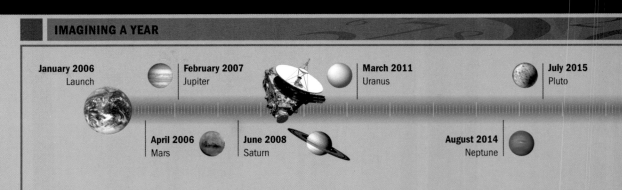

January 2006
Launch

February 2007
Jupiter

March 2011
Uranus

July 2015
Pluto

April 2006
Mars

June 2008
Saturn

August 2014
Neptune

top; it performs a complete circuit in 26,000 years. In addition, gravitational effects from Jupiter and Saturn cause the orbit to become more and less eccentric in a cycle of 413,000 years. At the moment we are about 1.5 per cent closer to the Sun in winter than in summer, but this is enough to make summer nearly 5 days longer than winter. Meanwhile the tilt of the Earth's axis (the "obliquity of the ecliptic") is now at just over 23 degrees, and rocks to and fro by about 2 degrees in a cycle of 41,000 years. At the moment the tilt is decreasing, which favours the onset of another ice age, but that will almost certainly be overcome by global warming.

Once around the Sun

Years vary greatly on the planets in our solar system. For Mercury the year is about 3 Earth months; Venus 7 months; Mars 23 months; Jupiter 12 years; Saturn 29 years; Uranus 84 years; and Neptune 165 years. So the closer a planet is to the Sun, the faster it whizzes round, in accordance with rules described by Johannes Kepler in 1619.

The years of all the planets are slowly growing longer. As it burns, the Sun emits a constant stream of charged particles – the "solar wind" that is responsible for phenomena such as the aurora borealis (above). By emitting these particles, the Sun is losing mass – nearly 7 million tonnes per second – and as its mass decreases, its gravitational pull weakens, which causes the planets' orbital periods to lengthen, little by little.

Earth's solar year
is becoming
1.5 microseconds
longer
per year

Twelve notches represent a full year

2029
Probe leaves solar system

NASA's *New Horizons* space probe was launched in 2006. Travelling at a speed of nearly 60,000 km/h (40,000 mph), it took a year to reach the orbit of Jupiter, and will leave the solar system 23 years after it took off. The probe is scheduled to make a close fly-past of Pluto in 2015.

CHAPTER 3

SETTING THE TIME

Short intervals of time are fairly easy to judge. You can tell when five seconds have passed, or five minutes, or an hour, even though you may be misled by circumstances into slightly over- or underestimating. On the other hand, you are unlikely to be able to judge intervals much shorter than a second, since our bodies are not designed to perceive such fragments of time; and somehow human "time-sensing" disappears for intervals longer than a day. For an event even a few days in the past you usually have to line it up with other roughly contemporary events to work out when it happened – for example, "When did that parcel arrive? It was the day after I went shopping, so it must have been Tuesday – five days ago."

To measure short intervals, we use clocks and electronic timers, which measure everything from the boiling of an egg to an Olympic sprint. For those longer intervals, we use diaries, and especially calendars. The calendar may just look like a picture that you hang on the wall, but in fact it is a long-interval time-reckoner; it works for both past and future; and it has an immense history.

MARKING MIDWINTER

MIDWINTER HAS LONG BEEN OBSERVED by countless cultures around the temperate world. The Sun is weak and low in the sky, the day is short, and the weather is still getting colder. For hundreds of generations, people have held festivals of all kinds to try to persuade the gods to bring back a higher, stronger Sun and warmer, longer days. Many such festivals were eventually incorporated into religious ceremonies, including Christmas.

Germanic peoples used to celebrate a pagan festival of Yule or Yule-tide, which may have developed from the Roman Saturnalia, but was later absorbed into Christmas. The Yule log was originally an entire tree, carefully chosen and brought into the house to provide warmth. Later it came to mean just a large log, of which one end would be placed in the fire and the rest slowly pushed in.

Ynglinga saga, written about 1225 in Old Norse by the Icelandic poet Snorri Sturluson, mentions a Yule festival in AD 840. The participants slaughtered several animals for a feast, and drank copious quantities of ale; later King Haakon I decreed that "everyone… had to keep the holiday while the ale lasted". He sounds like an excellent ruler.

BELOW: In Scandinavia, midwinter solstice is marked with a bonfire. The Yule log at Christmas is part of the same winter tradition.

The Chinese celebrate the Dongzhi festival: families get together and make *tangyuan*, which are brightly coloured balls of sticky rice that symbolize reunion. In Taiwan some of these *tangyuan* are attached to doors, windows, and tables to keep evil spirits away from the children. Zuni and Hopi Native Americans in New Mexico celebrate Soyal to mark the new cycle of the wheel of the year, and to bring the Sun back.

In what is now Iran, the ancient Persians celebrated Shab-e Yalda on the longest night of the year, in honour of the birth of Mithra, their god of light and truth. Since the rise of Islam, however, Shab-e Yalda has become a gentler family occasion, but fresh fruit is still served to help persuade the gods to look after the crops.

The winter solstice can be calculated to the second by astronomers, but for ordinary people it falls on 21 or 22 December

BELOW: Bolivian Indians make sacrifices to pre-Columbian gods on midwinter's eve, in the Alto Plano of the Bolivian Andes.

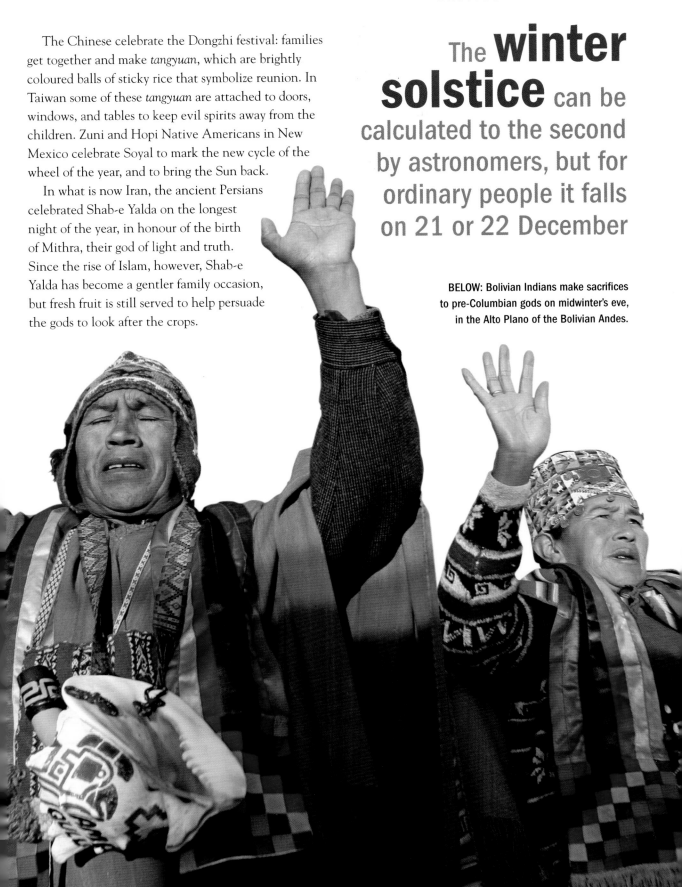

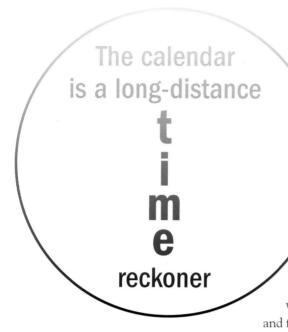

The calendar
is a long-distance
t
i
m
e
reckoner

All these celebrations and festivals were perhaps the beginning of calendars, for it must have been in the interests of anyone in a position of authority to know how the year was progressing, and how long it would be until the weather improved.

When is midwinter?

Technically, the midwinter solstice is the moment when the Earth's axis is at its maximum angle from the axis of the rotation of the solar system. The solstice is calculated to the second by astronomers, but for ordinary people generally refers to the shortest day of the year, which in the northern hemisphere is 21 or 22 December, and in the southern hemisphere 20 or 21 June. After the winter solstice, the Sun begins to rise earlier day by day, and farther along the horizon to the north (in the northern hemisphere), and the days start getting longer.

Several thousand years ago, Stone Age people in many countries built great monuments to mark the midwinter sunrise. Newgrange tomb in Ireland is a vast low-domed building constructed around 3000 BC. Four minutes after sunrise on midwinter's day, the Sun shines through a slit in the roof – or "roof-box" – and 19m (62ft) along a passage, to light up the floor of the central burial chamber for some 17 minutes. The first flash of sunlight would probably have appeared exactly at sunrise when the tomb was built.

There is a similar though rather smaller building at Maeshowe on the island of Orkney in Scotland: it is a chambered cairn and passage grave, with a long, narrow tunnel entrance exactly aligned with the midwinter sunrise.

BELOW: Newgrange tomb in County Meath, Ireland, is a masterpiece of Celtic astronomical architecture.

LEFT: The Castlerigg stone circle near Keswick in northern England was built around 5000 years ago in a commanding position with views down two valleys. Among the 38 stones are alignments marking both summer and winter solstices.

Both these buildings had tombs near the centre, and it seems likely that the most important members of the community were buried there – perhaps the local chief and his family at the time the building was completed. This in turn reinforces the idea that the people involved had some notion of an afterlife, or at least wanted in death to get close to their gods – and midwinter's day was clearly important in the spiritual order of things.

Chinese archaeologists have discovered a midwinter sunrise observatory at the Taosi site in Shanxi Province. Built about 4000 years ago, it seems to have been designed to mark the sunrise over the Chongfen mountain.

In Egypt, the mortuary temple of Queen Hatshepsut at Deir el-Bahari is equally remarkable. Built around 3450 years ago, the temple has three tiers, behind the

top of which is an open court that leads through a doorway into a chapel carved from the bedrock. The back of the chapel is about 43m (141ft) from the entrance to the court. Just after 6.30am on midwinter's day, the rising Sun sends a shaft of light straight through to a niche in the wall of the chapel, and the patch of sunlight moves slowly across the wall for 11 minutes before it disappears.

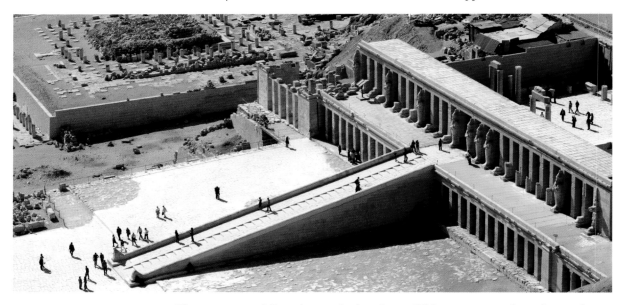

ABOVE: The temple of Hatshepsut lies on the west bank at Luxor. The midwinter sunlight entering the chapel underlined the power and mystery of the all-powerful sun-god, Amun-Ra.

OPPOSITE: Stonehenge on midwinter's eve has a mythic atmosphere. Although New Age travellers flock to the monument on midsummer's eve, they may be six months late, as the site's main purpose may have been to mark midwinter.

The main axis of Stonehenge, built at least 4000 years ago, is aligned towards the midwinter sunset (rather than sunrise), and the Great Trilithon was built "outwards" – with its smooth side facing towards the setting Sun. Even though this appears to be "back to front", and even though modern disciples and druids flock to Stonehenge at the other end of the year – to celebrate the midsummer sunrise – one of the purposes of the huge monument was clearly to mark midwinter.

Many people today marvel at the skill and mathematical ability of those Stone Age people in calculating the dates, and planning such monuments to mark the midwinter sunrise and similar events. But without artificial light, books, computers, or TV, there would have been little to do on long winter nights. Some people probably told stories or sang songs. Others must have enjoyed watching the dark sky and observing the pageant of Moon and stars sailing across it, not to mention the curious paths of the planets – the wanderers, as the Greeks came to call them – until the first light of dawn drove the stars from the sky and the first flaming edge of the Sun came into view.

Watch for a few days and you will see that the Sun does not always come up or set at the same points on the horizon: in fact, the positions of both sunrise and sunset move a little every day. In the northern hemisphere they reach their most northern points at the midsummer solstice, on or near 21 June; then they move a little to the right every morning, and to the left in the evening, until they arrive at their most southern points at the midwinter solstice, after which they begin to move back again. At both spring and autumn equinoxes, the Sun rises due east, and sets due west for any observer with a level horizon.

To track the motion of the sunset or sunrise along the horizon is easy. I am an early riser, and would rather watch for the sunrise. You might find sunset times more agreeable to track. The method is essentially the same: just look for the last flash of sunset instead of the first flash of sunrise.

Tracking the sunset or sunrise

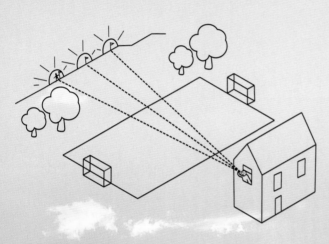

ABOVE: Find a fixed point from which to observe the sunset.

CHOOSE A FIXED VIEWPOINT – a rock to stand on, or a convenient window to look out of – and persuade a friend to stand on the horizon with a stick or other marker (with the date written on it) to plant in the ground. This is simple if you have a clear and close horizon, but impossible if there are woods in the way.

When you see the first flash as the Sun rises, signal or shout to your assistant to move left or right and plant the marker exactly in line with the Sun. The more distant the horizon, the more difficult the signalling becomes – you may need to use a mobile phone – but the more precise your alignment will be.

If you have the time and the patience, mark the sunrise position every day, but if not, once a week will demonstrate the pattern. When you get close to 21 June or 20 December, try to mark the position every day, to get exact directions for the solstice.

Those who live in a city may be able to track the sunrise without assistance, by noting where the first flash of sunlight appears – over the corner of a building, or beside the end of a bridge, through an arch, and so on. Write these notes down each day and you have a record of the moving sunrise. With a skyline varying in height, the changes in position from day to day are unlikely to be regular in distance, but they will form a progression, and you should be able to discover the extreme positions of sunrise, and therefore determine the day of the solstice and the length of the year (even though it might be easier to look these up, since they have been measured before).

Using this method, Stone Age people could easily have discovered the direction of midwinter sunrise, and checked it for several years, perhaps marking the alignment with posts driven into the ground. Then they might have decided to immortalize the alignment by building a vast monument – such as Newgrange tomb – to mark the turn of the year, and to be prepared to celebrate the correct day in future years. Building the monument was a colossal undertaking, but getting the alignment right required no great mathematical or scientific ability: merely curiosity and patience.

Once any such monument has been erected, or even wooden posts as markers, counting the number of days in the year becomes possible, although tedious. You could scratch a mark on a wall or a bone every day, or have a collection of 400 pebbles in one hole or pot and every morning move one to another container, but it would be easy to miss one, or to move two by mistake, and if you went away on a hunting trip you might not trust anyone else to do the job correctly.

BELOW: Tracking sunset and sunrise is easy with a city skyline.

The earliest known example of an effective annual marker was nothing to do with the Sun – at least not directly. In Egypt, 95 per cent of the land is desert, and for the last 7000 years the Nile valley has been the only fertile and habitable part of the country.

THE EGYPTIAN METHOD

ABOVE: The Nile at Luxor, in Upper Egypt. The river has not flooded seriously since the 1960s when the Aswan High Dam was built in the south.

THE NILE RISES FAR TO THE SOUTH, in Ethiopia and Uganda, and flows from south to north right through Egypt. Every sweltering summer it used to flood, bringing a tide of wonderful rich black alluvial soil, and there were just three seasons: flood, growth, and harvest.

The remarkable thing is that the flooding came with great regularity, year after year; so the day of the flood was an effective annual marker. No need to build a monument, or even work out an alignment: the Nile would reveal the start of the year. In order to measure the depth of the flood, nilometers were built – graduated columns of stone in pits with channels to allow the water in from the river. The depth of the water could be read from the graduation marks on the column, and one result is that there is a continuous record of the maximum depth of the water for the last 1400 years.

For most people, all that mattered was that the flood had come: they could celebrate the New Year, beginning with the Egyptian month of Thoth. For the priests, however, this regularity meant power. They could easily note the maximum depth of the water on the nilometer, and then count the days until the next maximum. That gave them a measure of the cycle of the seasons – the length of the year. Their first estimate was a year of 360 days, which they divided into 12 months of 30 days. With this knowledge they could not only make predictions, but also lay down timetables and regulations for the whole year.

Then a new marker became apparent. Egyptian astronomers noticed that the Nile flooded on or very close to the day on which Sirius, the "Dog Star", rose in the dawn sky just ahead of the Sun. Sirius is the brightest star in the sky; hence its name, from the Greek (Seirios) meaning scorcher. The astronomers used the appearance of Sirius as a more precise marker of the beginning of the year, realized the year was more like 365 days than 360, and added five extra days of festival – the feast days of Horus, Isis, Nephthys, Osiris, and Set – to bring their year more into line with the heavens.

In due course, the astronomers discovered that the year is not exactly 365 days, but about six hours longer. They suggested adding a quarter of a day to the year. The priests, however, said the calendar was sacred, and could not be changed, and it was not until 30 BC when the Romans moved in that the Egyptian calendar was improved by the addition of an extra day every fourth year. In fact the number of days in the year is approximately 365.24219, or 365 days 5hrs 48mins 45.2secs, or just less than 365¼. Nevertheless, 365¼ is good enough for most purposes.

ABOVE: Sirius, the Dog Star, rises over the Western Desert. Its appearance ushered in Egypt's annual renewal through the flooding of the Nile.

The dog st★r heralds the flooding of the Nile

THE CALENDAR AT KOM OMBO

Kom Ombo is an agricultural town on the Nile towards the south of Egypt. It was originally called Nubt - the city of gold. The splendid double temple, built around 2200 years ago, is dedicated to both the crocodile god Sobek and the falcon god Horus. On a wall of the temple is carved a calendar showing New Year's Day on 19 July when Sirius rose, the days on which festivals were to be held, and what kind of offerings were due.

The enthusiastic sunrise watcher does not have to stop at the midwinter and midsummer sunrise positions. Many sets of markers constructed in ancient times have been discovered in North and South America. A striking example is the 300m (980ft) row of 13 towers at Chankillo on the northern coast of Peru, built around 300 BC.

DIVIDING THE YEAR

THERE IS EVIDENCE OF A SUBSTANTIAL SETTLEMENT at Chankillo with areas for festivities, and a clear observation point – a doorway in a high wall. From this doorway the eastern horizon comprises a low ridge running roughly north–south, about 235m (771ft) from the observation point.

Along this ridge are 13 chunky stone towers, ranging in height from 2m (6.5ft) at the highest part of the ridge to 6m (20ft) at the lowest, thus forming a castellated horizon. The midwinter sunrise appeared just to the left of the left-hand tower, in the narrow V between it and the shoulder of a distant mountain. The midsummer sunrise appeared over the tower at the other end of the row.

The gaps between the central towers are only 5m (16ft) wide, which means that the sunrise would appear in each gap for only a few days, and then there would be a 10-day wait until it appeared in the next gap. This suggests that the natives may have used a 10-day "week". There is a second observation point on the other side of the ridge, from where equivalent sunsets must have been viewable between towers.

Another apparently simple way to slice the year into equal portions would be by Moons. Once the first markers were down it would be a simple matter to count the number of new Moons between midwinter and midsummer: there are usually six or seven. Mark the position of the sunrise immediately following each new Moon as the

NICOLAS ORESME

Nicolas Oresme was born near Caen in about 1323, and became Bishop of Lisieux. In his *Book of Heaven and Earth* he discussed the apparent motion of the heavenly bodies, and concluded that we should see exactly the same thing if Earth were spinning, while they stayed still – and that this would be much more economical than the entire Universe revolving. He dismissed the notion that a spinning Earth would cause a great wind, since he said that earth, water, and air would all move together. But he worried that the Sun and Moon do not seem to be synchronized, and wondered whether they actually tell the same time.

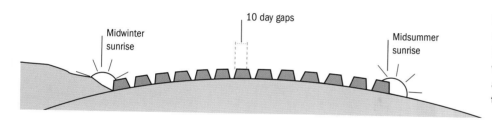

LEFT: The midwinter sunrise appears just to the left of the left-hand tower at Chankillo, while the midsummer sunrise appears just to the right of the right-hand tower.

sunrise moves gradually south from the midwinter point. The first day the sunrise appears between one of the markers could be called the beginning of a new Moon or month. The difficulty with this system is that the Sun and Moon are not in sync. The Moon takes 29.53 days for one lunation (new Moon to new Moon), which means that 12 lunations take 354.36 days, and 13 lunations take 383.89 days, while the year is 365.24 days, or about 12.37 lunations.

In his 14th-century work *De Proportionibus Proportionum*, Nicolas Oresme (see opposite page) discussed whether the celestial motions are commensurable: is there a basic interval of time that is a common factor of the orbital periods? He said he might have expected God to create such a unit, but concluded that there is no such unit, that no two periods are commensurable, and asked whether time as measured by the Sun and the Moon is the same time.

Calendar months

Measuring the length of the year was difficult for the ancients, because apart from the flooding of the Nile, the markers were subtle. The cycles of the Moon, however, are much easier to follow. The dark Moon, and the first crescent of the new Moon, are easy to spot, unless there are thick clouds; so anyone can count the number of days from one dark Moon to the next, which is called one lunation. Even if it is cloudy for several days you can always wait, and count the following month. Counting from one full Moon to the next will do, although the Moon can look full for three or four days; so you have to be careful, unless you get a very clear sight. The fact that counting lunations is easy meant that Moon calendars were probably in use long before anyone counted the length of the year.

In Bali the people use a calendar based on Sun and Moon, but also a permutational one consisting of ten cycles of various numbers of days. Whenever the cycles converge there is a holiday. These cycles do not actually measure the passing of time, but simply provide a set of markers of propitious days for doing business or starting to build a house, or meeting a new person, and above all an excuse to celebrate.

One lunation, the interval between
dark Moons
is... 29.53 days

Marshack was born in the Bronx in New York. In his forties he became interested in the early history of humans, and began researching Palaeolithic artefacts. His book *The Roots of Civilization* (1972) tried to show that our ancestors of 30,000 years ago thought in the same way as we do today. In it, he started with the Ishango bone, saying that his first assumption was to ask whether the series of odd and different counts on the bone "could be related somehow to a time count and perhaps to a lunar count". In other words, rather than start with an open mind, he jumped to the conclusion that this was a Moon-based calendar, and then spent the rest of the book attempting to justify it. Some of his ideas were attacked, but his work established the existence of early calendars, and according to his obituary in *The Times* newspaper, "had an immediate effect on archaeological thought".

19   +   17   +   13   +   11 = 60

ABOVE: Is the Ishango bone (both back and front shown here) a 25,000-year-old lunar calendar?

Stone Age calendars

In a cave in the Dordogne valley in France was found a 30,000-year-old 10cm (4in) fragment of bone from an eagle's wing. On the surface was carved a serpentine pattern of notches, which at first was taken to be merely decorative. However, independent American archaeologist Alexander Marshack examined the notches under a microscope, and realized that they are in groups of 14 or 15, so that there are 29 or 30 notches in each row. He suggested that this is the period between two lunations (ie full Moons), and therefore that the bone might be a Palaeolithic calendar.

The Ishango bone, found in 1960 in what was then the Belgian Congo just before it became the Democratic Republic of the Congo, is the fibula (leg bone) of a baboon that died 25,000 years ago. This appears also to be a lunar calendar: the notches were carved in groups, growing gradually bigger and smaller along the length of the bone, in a way that seems to correspond to the phases of the Moon – increasing notches to suggest waxing and shrinking notches for waning.

There is speculation that these lunar calendar bones were carved for women to keep track of their menstrual cycles and therefore their fertility, and they would probably have been useful even for those whose cycles do not match the Moon's.

The notches on the **Ishango** **B**on**E** may mean that someone was marking the phases of the Moon 25,000 years ago

Another possibility is that some bored youth, lacking TV, mobile phone, and computer, watched the night sky and the changing shape of the Moon, and decided for fun to keep a record of it, to see whether the pattern was regular. Having done that, he or she could then have used the knowledge to make simple predictions: "In three days the Moon will be full and round." Possession of such knowledge might well have raised his or her status in the family or group.

Just to the west of Kiev in Ukraine is what used to be the Gontzi mammoth hunting ground, and the tip of a mammoth tusk discovered there seems to have survived from just after the end of the last ice age, some 20,000 years ago. It is inscribed around the edge with a series of notches which can be interpreted as a lunar calendar that covers a period of four months – long enough to cover an entire season, and in three cycles a whole year.

ABOVE: Calendars have been found carved on mammoth bones in Ukraine, in periods covering a whole season.

LEFT: Palaeolithic youths, during the long winter nights, may have taken far greater note of lunar patterns than we do today.

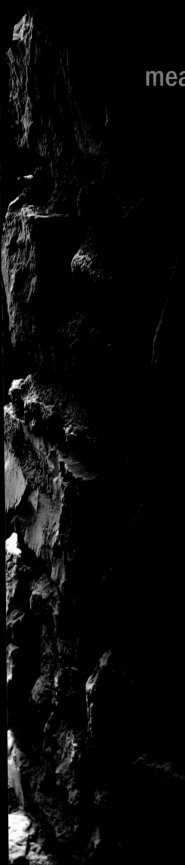

ABOVE: Betelgeuse (top left), Orion's brightest star, is visible in the northern hemisphere between September and May.

In a cave in the Ach valley in Germany, archaeologists found a 35,000-year-old fragment of ivory from the tusk of a mammoth. On one side is carved a human figure with arms and legs outstretched. This is one of the earliest known representations of the human figure, but the precise proportions of the figure suggest that it might possibly represent the constellation of Orion. On the other side is a sequence of 86 notches, and it happens that 86 is the number of days for which Betelgeuse, one of Orion's prominent stars, is visible in the northern hemisphere.

Whatever the reasons for the carving, these bones are arguably the oldest known calendars in the world. Thirty-five thousand years ago people were still hunter-gatherers – agriculture had not been invented – so they would not have wanted a calendar to tell them when to plant crops. But perhaps a calendar might have allowed them to make better predictions of when to expect the appearance or migrations of the animals they hunted, and the berries and nuts they gathered.

These calendars were probably for personal use: we can guess that whoever carved them used them himself or herself, or for the benefit of the family or group. There came a time, however, when human beings were gathered into larger groups, or tribes, or perhaps even towns, when those in charge wanted to have control over some of the behaviour of the rest. There must have been advantages in being able to tell people when to come together for meetings or festivals, and when to pray and when to pay their tithes or taxes. Hence the public display of the calendar at Kom Ombo (see p123).

THE METONIC CYCLE

The ancient Greeks were expert astronomers, and understood the problems of calendars based on either Sun or Moon alone. In 432 BC, Meton of Athens worked out that in 19 years, as measured by the Sun, there is the same total number of days as in 235 months, as measured by the Moon. In other words, he could reconcile Sun and Moon times by combining seven 13-month years with twelve 12-month years. Seven 13-month years have a total of 7 x 13 x 29.53 = 2687.2 days; twelve 12-month years have 12 x 12 x 29.53 = 4252.3 days. Adding the two makes 6940 days (to the nearest day), while 19 years of 365.25 days also make a total of 6940 days. In practice, the difference between the two is only 2 hours. The Metonic cycle could be achieved in practice by having 12-month years, and inserting an extra month in (for example) years 2, 5, 8, 10, 13, 16, and 19; then repeating this sequence. Sadly for Meton, his cycle proved too complicated to be adopted at the time, although even today it is useful for planning space missions, as the basis of the Hebrew calendar, and also for calculating the date of Easter.

6940 days =
235 months =
19 years =
1 Metonic cycle

BELOW: Hunter-gatherers did not need calendars for planting and crops, but they may have needed them for migrations of animals and seasonal foods.

The positions of moonrise and moonset on the horizon vary from day to day – like the positions of sunrise – and from year to year. The time when they are farthest apart – when the Moon travels farthest along the horizon in one leap, technically known as the maximum range of declination – is called the "major standstill", and it happens every 18.61 years. The next one is due in the spring of 2023.

Lunar standstills

THE EXTRAORDINARY COLLECTION of megalithic arrays erected some 5000 years ago at Callanish on the island of Lewis, to the west of Scotland, seems to have been built to celebrate the major standstill. There are ten stone circles, five smaller stone groups, seven rows of stones, and nine single standing stones, all within a small area of what is mainly peat moorland. Some sites point to the midwinter sunset, but most appear to be focused on the major standstill. The main arena is an avenue of stones leading to a dramatic circle of white stones, looking almost like giant teeth reaching up towards the sky.

At the major standstill, the full Moon rises (as seen from the main circle) out of the "Sleeping Beauty" (also called Cailleach na Mointeach, or Old Woman of the Moors), which is a mountain on the horizon that has roughly

BELOW: Standing stones at Callanish on the Isle of Lewis appear to celebrate the major standstill.

the shape of a woman lying on her back. The Moon then appears to roll around the horizon, for at the latitude of Callanish it does not rise high in the sky, and then sets in the Clisham Hills, or between the stones of the small central circle, as viewed from the far end of the dramatic avenue of gaunt white stones. In addition, the midwinter Sun sets by "rolling down" the face and forehead of the Sleeping Beauty, slipping behind the next hill, and then briefly flashing again through a small valley.

There are two remarkable features of this phenomenon, quite apart from the fact that Callanish is remote from any large centres of population, and a great workforce must have been needed to find, transport, and erect the stones. First, this avenue and small circle is only one of some 20 groups of stones spread over 5 sq km (2 sq miles), but all within line of sight of a platform on the side of a hill – perhaps a sort of "mission control" for any ceremonies that were performed.

Second, the cycle takes 18 years, which was probably not far short of the average life span when the stones were erected. To track it precisely would have needed several cycles, so the information must have been passed down the generations – or some unusually old person persuaded the young ones to take them seriously.

Lunar standstills may also account for various alignments in the United States. At the Hopewell site in Newark, Ohio, the main axis of the circle and octagon earthworks is aligned with the major standstill moonrise. From the Great House Pueblo at Chimney Rock in Colorado, two rock towers frame the same moonrise. Both major and minor standstills (when moonrise and moonset are closest together) are features of the Sun Dagger site on Fajada Butte in Chaco Canyon.

ABOVE: The time when the moonrise and moonset are farthest apart on the horizon is known as the "major standstill".

We have no idea what happened at the stones of Callanish or Stonehenge, since whoever lived and danced or worshipped or feasted there left no records apart from the stones themselves. The ancient Egyptians, however, were kinder to us, for they left all sorts of records in the shape of writing and paintings, and we have at least some idea of what they were doing – or at least what they claim they were doing. Even more explicit were the Romans.

THE JULIAN CALENDAR

THE EGYPTIAN PRIESTS ESTABLISHED THEIR CALENDAR as a religious object and rulebook; that is, the calendar itself was sacred, and it defined the rules which the people had to obey. It was therefore an instrument of power for the priests. Secular rulers realized that they too would benefit from such a device, and one of the first to do so was Julius Caesar, who was essentially a brilliant general with a vast empire to control, and he had the sense to realize that controlling people's time was crucial.

According to legend, the first Roman calendar was established by Romulus, who together with his brother Remus had been reared by wolves, and went on to found the city of Rome (named after Romulus) in 735 BC. From then on, Roman years were numbered *ab urbe condita* or AUC (from the founding of the city).

Romulus was keen on the number 10, and his year had only 10 months. Some of the early months were named after gods – Mars (god of war), Aprilis (same as Aphrodite and Venus, goddess of love

RIGHT: Romulus and Remus, suckling on the she-wolf, were the legendary founders of Rome, from which event Roman time was dated.

Romulus was keen on the number 10 and his year had 10 months

JULIUS CAESAR

Gaius Julius Caesar (100–44 BC) was born into an administrative family: his father was governor of the province of Asia. After an exciting early life, which included being captured by pirates, he joined the army, became a successful general, conquered Gaul (France), and even invaded England in 55 BC. This military success gave him great political clout, and after many shenanigans he became dictator in 48 BC. On a trip to Alexandria he became infatuated with Cleopatra (who was proud of the calendar devised by Claudius Ptolemy) and they remained lovers for 14 years. On 15 March ("the ides of March") 44 BC, Caesar was assassinated in the senate by a group of men, who between them delivered 23 stab wounds.

and beauty), May after Maia (goddess of growth), and Juno (queen of gods) – and the last four were just numbered septem (7), octo (8), novem (9), and decem (10). Around 700 BC, King Numa added two months at the end of the year, calling them Januarius and Februarius.

BELOW: A Roman calendar such as this would have been on public display in the forum of all Roman towns and cities.

A new start

By the time Julius Caesar came along, the calendar had drifted two months away from the solar year, and he decided it needed sorting out. He hired the most able philosophers and mathematicians to organize it logically, and in 46 BC (709 AUC) introduced the Julian calendar, establishing 25 December as the date of the winter solstice. To get back into line, he had to add a couple of months that year, which came to be called "the year of confusion" since there was much argument about what taxes, interest, and bills had to be paid. His old political rival Cicero complained that Caesar wanted to rule not only the Earth but also the stars.

Caesar also decreed that the year should begin with January, rather than March. This unfortunately meant that the last four months of the year were wrongly named – September became not the seventh but the ninth month, and so on – but we have now been stuck with those wrong names for 2000 years, and they probably won't change.

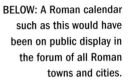

The leap year made Caesar's the most accurate calendar that the world had ever known

The Julian calendar had 12 months, and every four years a leap year, in which an extra day (a leap day) was added to the month of February, to give it 29 days. Generations of children have learned this in the nursery rhyme, of which Groucho Marx said, "My favourite poem is the one that starts 'Thirty days hath September' because it actually tells you something":

> **Thirty days hath September, April, June, and November.**
> **All the rest have 31, except February alone,**
> **Which has but 28 days clear, and 29 in each leap year.**

This trick of the leap year made Caesar's the most accurate calendar that the world had ever known. The seventh month was eventually named after him (July) and the eighth month after his successor, Augustus.

Caesar's calendar made a considerable difference to ordinary Romans. Calendars were carved in stone and painted on walls, and became public knowledge. Instead of having a semi-mystical system of dates and festivals, under the dubious control of politicians and priests, there was suddenly a robust and public arrangement of days and dates, which allowed everyone to make proper plans throughout the year.

Constantine the Great

Three hundred years later, Constantine the Great (Flavius Valerius Aurelius Constantinus; AD 272–337) adopted Christianity, became Emperor, and decided to put religion back into time. He left most of the Julian calendar intact, but re-invented a seven-day week, partly because of the astrological significance of there being seven planets (which is probably why the Babylonians had used a seven-day week a thousand years earlier), and partly because the Bible said that God had rested on the seventh day, and Constantine thought that mere mortals should do likewise. The Emperor further decreed that this holy day should be the Sun's day, which was not popular with Jews or Roman pagans, who had always celebrated Saturn's day as their day off.

OPPOSITE: The Romans celebrated the festival of Cerealia in mid-April by the Julian calendar. It lasted seven days, but this painting, *Spring*, took artist Sir Lawrence Alma-Tadema four years to complete (1890–94).

A LEAP-FREE CALENDAR

In the Persian Jalali calendar, adopted in 1079 on the recommendation of the mathematician and astronomer Omar Khayyám among others, the year began on the day of the spring equinox. The calendar was tricky to use and needed precise astronomical observations, but originally did not need leap years, although the months varied in length. A system of eight leap days in every cycle of 33 years was later introduced.

The last major change in the calendar came on 24 February 1582 when Ugo Buoncompagni, better known as Pope Gregory XIII, signed the papal bull that ratified the tweaks to Constantine's system suggested by a Calabrian doctor, Aloysius Lilius, and a tubby Jesuit astronomer called Christopher Clavius.

THE GREGORIAN CALENDAR

ABOVE: Pope Gregory XIII's tomb in the Vatican commemorates his introduction of the modern calendar.

THE JULIAN CALENDAR WAS BASED on the assumption that one year is exactly 365.25 days, which is in fact about 10¾ minutes too long. By 1582 the accumulated error was 10 days, which meant that the spring equinox had already moved to 11 March, and was creeping even earlier. This had led to a steady sliding of the date for Easter, the most important date in the Christian calendar, which had been set as the first Sunday after the first full Moon after the spring equinox, and the Catholic Church did not approve of a creeping Easter.

Astronomers and mathematicians realized what the problem was, but they had to persuade everyone to shorten the year by 10¾ minutes, which was far from trivial. However, Gregory's team devised a cunning plan. Clearly it was no good hoping to get people to lose a few minutes every year: it had to be a whole day or two every so often. They worked out that 10¾ minutes is 1/134 of a day, and therefore rather than cutting 10¾ minutes each year they could cut one whole day every 134 years. Hmmm, not convincing. How about two days every 268 years? No better. But three days every 402 years was more promising. Why not call it three days every 400 years? This allowed them to tinker with the leap years, and introduce a simple rule that was much more accurate than Caesar's:

Leap years are all the years whose number is a multiple of four, except those century years that are not multiples of 400.

RIGHT: Traditionally, women may propose to men only on 29 February, the additional leap year day.

An unpopular leap

And the scheme worked: we have been using this calendar ever since. But Gregory also had to get the year back into sync with reality. It had drifted so far ahead of the Sun that he and his team decided they simply had to lose 10 days. Lilius suggested losing these days over 40 years by just not having leap years, which would not affect people much, but Clavius said they should do it all at once, and Gregory was persuaded. So the 1582 calendar moved from Thursday 4 October directly to Friday 15 October.

This was not popular. In Frankfurt, people rioted because their days had been stolen, and all sorts of people were upset because they had lost saints' days, birthdays, and festivals, or because their taxes or rent had not been paid. The confusion was made worse by the fact that the Catholic Church was no longer all-powerful, and a great many Protestants, not to mention Jews and Muslims, simply paid no attention to the Pope's edict.

In the end, the new Gregorian calendar was adopted, mainly because it was much more logical than the old one, and now it is used throughout the world. However, in Britain and the British Empire (including the east of North America) the change was not made until 1752, by which time 11 days had to be lost to get into sync, and Wednesday 2 September was followed by Thursday 14 September. Again there was some public outrage, with people demanding the return of their 11 days, and several died in riots in Bristol.

ABOVE: The German mathematician and astronomer Christopher Clavius was one of the main architects behind the Gregorian calendar.

In October 1582 Pope Gregory XIII cut **10 days** off the calendar

WHEN THE YEAR IS LEAPED

After Gregory's bull of 1582, the years 1584, 1588, 1592, 1596, and 1600 were leap years, but 1700 was not a leap year, because 1700 is not a multiple of 400 (17 is not a multiple of 4). This made the effective length of the year 365.2425 days (or 365 days, 5hrs, 49mins, 12secs) rather than Caesar's 365.25 days, and therefore very close to the actual value of 365.24219 days – only 27 seconds out of sync with the Sun.

Caesar's year = *365.25 days*
Gregory XIII's year = *365.2425 days*
Actual year = *365.24219 days*

ABOVE: William Hogarth's 1755 painting *An Election Entertainment* shows a banner reading "Give us our Eleven days" lying on the floor of the tavern (bottom right), dropped by a Tory party supporter.

The satirical painter William Hogarth reflected the mood in his famous series of paintings of the 1754 election campaign. A popular slogan among the anti-reformers ran:

In seventeen hundred and fifty-three
The style it was changed to popery

Curiously the British tax year, which used to begin on 25 March (which was New Year's Day in the "old style" Julian calendar), carried on with the Gregorian calendar. The tax authorities refused to lose 11 days' worth of taxation, and therefore decreed that the tax year should be extended by 11 days, from 25 March to 5 April. When a leap day was skipped in 1800, the date was moved to 6 April, which is still the beginning of the British tax year.

One thing the new calendar failed to settle was the date of Easter, for which various sets of rules have been devised over the centuries and throughout Christendom, since the First Council of Nicaea convened by Constantine in AD 325. The process of calculating the date is called *computus* (Latin for calculation), and it is complicated by the fact that, although Easter Day is the first Sunday after the first full Moon after the spring equinox, there has been endless argument over whether that is a real full Moon (as observed) or a theoretical one, and which is the actual day of the equinox, and what happens when the full Moon falls on the day of the equinox, or on a Sunday.

ABOVE: The method for calculating Easter known as *computus* has remained the same since the Middle Ages.

One of the scholars involved in the *computus* was Dionysius Exiguus (Dennis the Small), who in AD 525 introduced the idea of dating the year from the time of Jesus, and calling it *Anno Domini* (Latin for "the year of Our Lord") or AD for short. He used the idea only for the calculation of the date of Easter, but it became incorporated into the Gregorian calendar, and has survived ever since, even among many non-Christians, along with the term "Before Christ" or BC, for earlier dates. A curious feature is that there was no year zero: in this system AD 1 followed 1 BC. In fact, all that Dionysius said was that he was writing during the consulship of Probus Junior, which was 525 years "since the incarnation of our Lord Jesus Christ" – so it is not clear whether he was talking about Jesus's birth year. Meanwhile, other evidence suggests that Jesus was actually born before 4 BC, and possibly as early as 18 BC.

Easter is the first Sunday after the first full Moon after the spring equinox

The complexities of the Gregorian calendar have prompted various people to suggest other possibilities. Moses Bruine Cotsworth, who was born in England in 1859 and worked as a statistician for the North Eastern Railway Company at York, realized that monthly accounting was complicated by the fact that the year does not divide evenly into weeks. He proposed a splendidly simple solution, which became known as the International Fixed Calendar.

A RATIONAL CALENDAR

ABOVE: The Islamic calendar, and the dates of its religious festivals, are based entirely on the phases of the Moon.

BELOW: George Eastman, founder of the Kodak Company, uniquely adopted Cotsworth's calendar for his business.

COTSWORTH'S YEAR WOULD COMPRISE 13 MONTHS, each of exactly 28 days. The extra month, called Sol, would come between June and July. Each month would start on a Sunday and finish on a Saturday, which means that 4 July, for example, would always be on a Wednesday, as would the 4th, 11th, 18th, and 25th of every month (except 25 December).

Because 13 x 28 = 364, these 13 months would leave the year one day short. Cotsworth proposed that Christmas Day should be on Sunday 22 December. As a holiday it would not count as a day, so 23 December would also be Sunday. (Alternatively the extra day could be called "year-day" and would fall between 28 December and 1 January.) When a leap year was needed, there would be an extra Sunday on 29 June. Easter would always be on 15 April.

He published his ideas in 1902 in *The Rational Almanac*, a tall thin volume of 472 pages filled with powerful arguments in favour of his system, but perhaps undermined by the inclusion of "180 illustrations explaining the Mystery of the Pyramids, Sphinx, Obelisks, Druidical Circles, Mounds, Vertical Stones etc."

Cotsworth moved to Vancouver, Canada, in 1910, and devoted much of his time and his considerable fortune to calendar reform. In 1922 he founded the International Fixed Calendar League. His scheme was endorsed by the Royal Society of Canada, and the League of Nations. He visited 60 countries in his efforts, but failed to bring about comprehensive change. He did, however, manage to persuade a lot of people, including George Eastman, founder of Kodak: for many decades, Kodak employees were paid in 13 monthly instalments.

The Islamic month begins with the first sighting of the new Moon

The Islamic calendar is used in Saudi Arabia and other countries around the Gulf and elsewhere for celebrating religious festivals and holy days. It is based entirely on the Moon. It has 12 months, one for each Moon of the year. According to the Quran, "Surely the number of months with Allah is twelve months in Allah's ordinance since the day when He created the heavens and the earth."

Each month of the Islamic calendar begins with the first appearance of the thin crescent after the dark Moon. When this appears can be calculated precisely by astronomers, and this calculation is good enough for most Muslims; some, however, especially in India, Bangladesh, and Pakistan, insist on actual observation of the new Moon. For many this means observation anywhere in the world, but for those who insist on a local sighting with the naked eye by a trustworthy man who can testify before Muslim leaders, there may be difficulties such as atmospheric pollution or cloud cover. In some cases, therefore, no one knows exactly when the month will start, which is a particular problem for the ninth month, Ramadan, when all Muslims are expected not to eat, drink, or have sex during daylight hours.

BELOW: Hundreds of thousands of Muslims gather at Mecca for Ramadan, the ninth month of the Muslim calendar and the time of fasting.

BELOW: Cairo at sunset, the beginning of the Islamic 24-hour day.

A few Gulf countries have adopted a simpler system: the month begins on the day after the 29th day of the previous month, as long as the Moon sets after the Sun, as seen from Mecca.

Islamic (and Jewish) days begin at sunset. Friday is known as "gathering day": at noon Muslims gather to worship at the mosque, and the day is a day of rest. The following day, Saturday, is normally regarded as the first day of the working week.

The problem of observing the first slim crescent of the new Moon was one of the spurs that drove Muslims to become world leaders in astronomy around a thousand years ago. Two other driving forces were the necessity of navigating deserts in the cool of the night, when the stars were the only guide, and the creation of the Bayt al-Hikma (House of Wisdom) in Baghdad by Caliph Harun al-Rashid (c.763–809). This was a library where scholars were paid to translate all the great works from Greece and India into Arabic, which was the form in which they came to be known in the West a few centuries later.

One of the major advances made by Islamic scientists and technicians around this time was the development of the astrolabe, an

ISLAMIC YEARS

Islamic years have only 12 months, which average 29.53 days (the time between new Moons). This makes a total of 354.4 days. In practice the months have 29 or 30 days, but the total is 354: this is 11 days shorter than the Gregorian calendar, so that Islamic years pass more quickly. They are counted since Mohammed moved from Mecca to Medina, which was on 16 July 622 AD and is counted as 1 AH (AH = *Anno Hegirae*). The year 2012 AD corresponds approximately to 1433 AH.

instrument that allowed the user to find the time, by day or night, from the positions of heavenly bodies.

More generally, however, observations of the Sun by day and of the stars by night gave astronomers more precise ways of measuring time. Determining the month is easy if you realize that during the year the Sun passes through all 12 constellations, as Vitruvius pointed out in the 1st century BC. When the Sun is in Taurus, for example, the month is June. In other words we do not need to note the passage of the seasons to follow the course of the year. How this knowledge led to the development of sundials and clocks is explained in chapter four.

ABOVE: The perfection of the astrolabe in the early medieval Islamic world allowed accurate observations of time.

The term "Mesoamerica" does not appear on many maps, but refers to a vast area in Central America and northern South America in which a variety of people and cultures appeared, thrived, and then disappeared, the last of these occurring after the Spanish invasion of the 16th century.

Mesoamerican calendars

THE OLDEST OF THE GREAT MESOAMERICAN PEOPLES were the Olmecs, who lived on the east coast of Mexico from about 1200 to 400 BC. The best known were the Maya, who established city states from about 250 BC, and the Aztecs, whose empire lasted from about AD 1300 until the Spaniards arrived.

These separate peoples spoke different languages but shared various cultural habits, and they all had elaborate calendar systems, frequently based on repeating cycles, since for them time was cyclical (see p47). Most of them used variants of a 260-day Mayan calendar, sometimes called the Tzolk'in calendar, which is still in use today in the highlands of Guatemala. There were 20 names (Imix, Ik, Ak'bal, and so on) and 13 numbers; so the first day would be 1Imix, the second 2Ik, then 3Ak'bal, and so on until 13Ben, which was followed by 1Ix, 2Men, and so on, until all 260 combinations had been used. This was a sacred calendar, and each day had its special signs and duties.

Names used in the Tzolk calendar: Imix, Ik, Akbul, Kan, Chikchan, Kimi, Manik, Lamat, Muluk, Ok, Chuwen, Eb, Ben, Ix, Men, Kib, Kaban, Etznab, Kawak, Ajaw

ABOVE: The Codex Cospi is from the Puebla-Tlaxcala area of Mexico and is a political cartoon mocking the Aztec authorities.

The Maya also used a 365-day calendar, which was called the Haab, consisting of 18 months of 20 days each, plus 5 unlucky days at the end of the year. Finally, their third calendar, the "long count", was simply the number of days that had passed since it started on 11 August 3113 BC, and was used to refer to any particular date in the past or the future. Unfortunately the long count "Great Cycle" lasts only 5130 years, and according to the Maya the world will end (or not; see p48) on 23 December 2012. Let us hope this turns out to be wrong.

The Aztecs also believed in cycles of time: every day had some religious significance, but because time was cyclical the day would come back. Like the Maya they used a 260-day calendar, which they called Tonalpohualli, which comprised 20 "weeks" of 13 days. This religious and festival calendar was used to work out the most propitious days for particular events. They also used a 365-day solar calendar called Xiuhpohualli, which was used for agricultural planning. These two calendars came into sync once every 52 years, which was called the "calendar round". The turn of this Aztec "century" was time for more festivities.

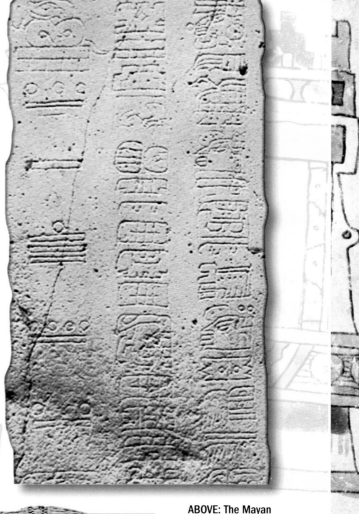

ABOVE: The Mayan stela known as La Mojarra dates from the 2nd century AD. The left column gives the long count date of AD 156.

AZTEC CALENDAR WHEELS

Both the Mayan and Aztec 260-day calendars were thought of as two wheels (one for name and one for number) which meshed together. Every day, both wheels moved one place, so that name and number changed. Each rotation of the small wheel represented a "week", and weeks 1, 6, 11, and 16 were particularly important. The two wheels came back into sync again after 260 days, when the cycle started again.

The Chinese lunisolar calendar has years that are approximately synchronized with the Sun, but months that are strictly synchronized with the Moon. The calendar runs in 12-year cycles, in which each year is named after an animal — so 2015 will be the year of the Sheep.

THE CHINESE CALENDAR

BELOW: Money is given in red envelopes, a lucky colour, at Chinese New Year.

EACH CHINESE MONTH BEGINS AT MIDNIGHT on the day of the dark Moon, and traditionally the Chinese New Year began on the second dark Moon after the winter solstice, which comes anywhere between late January and mid-February. New Year has been celebrated with firecrackers, feasting, and other festivities for at least 3500 years. Families gather together, sweep out the bad luck and old business with the dust, give presents, including money in red paper envelopes, display tangerines and oranges, and offer dried fruit to guests.

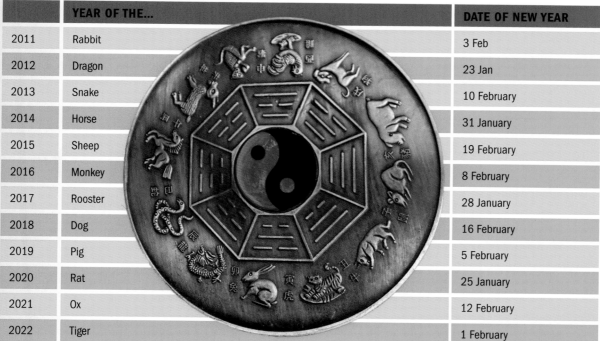

	YEAR OF THE...	DATE OF NEW YEAR
2011	Rabbit	3 Feb
2012	Dragon	23 Jan
2013	Snake	10 February
2014	Horse	31 January
2015	Sheep	19 February
2016	Monkey	8 February
2017	Rooster	28 January
2018	Dog	16 February
2019	Pig	5 February
2020	Rat	25 January
2021	Ox	12 February
2022	Tiger	1 February

In practice, most Chinese people use the Gregorian calendar, which is simpler because it is based on the Sun, but the traditional Chinese calendar is still used by Chinese people all over the world for festivals, and to choose auspicious dates for weddings, funerals, business deals, and other such events.

OPPOSITE: A Chinese dragon is intended to herald an auspicious New Year.

There is no obvious reason why there should be seven days in a week. The solar year is not an exact multiple of seven days; nor is the lunation. Yet a seven-day week is what we have.

THE DAYS OF THE WEEK

VARIOUS PEOPLES HAVE CHOSEN DIFFERENT "WEEKS". The French moved to a 10-day week (*décade*) in 1793, but reverted to seven days in 1802. The Soviet Union changed to a five-day week in 1929, and then to a six-day week in 1931, but went back to seven days in 1940. The Javanese people of Indonesia use a five-day week. The Hermetic Lunar Week was a plan by software developer Peter Meyer to divide the lunation into four quarters – new Moon, first quarter, full Moon, and last quarter, each of which might be six, seven, eight, or nine days, depending on the actual phase of the Moon – but this is too complex to be practical. The ancient Chinese and Egyptians used 10-day weeks, as did some Peruvians. The Maya and Aztecs used weeks of 13 days (see p144). In the end, seven days has turned out to be convenient, and is close to a quarter of a month.

Between

1929 and

1940

the Soviet Union changed the length of the week ~~once~~ ~~twice~~ three times

RIGHT: A French revolutionary calendar for 1793 reveals the 10-day week. The days were named *primidi*, *duodi*, *tridi*, *quartidi*, and so on.

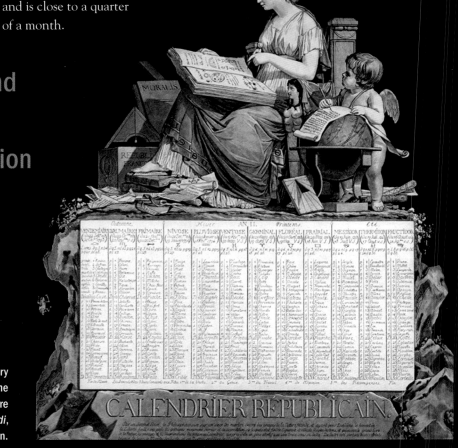

CALENDRIER RÉPUBLICAIN.

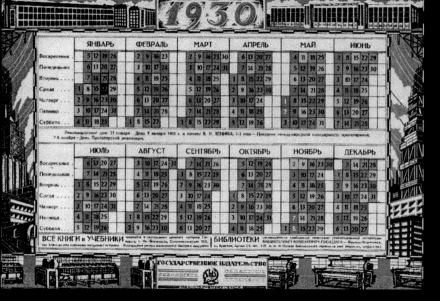

In the Western World the individual days are named after gods, originally in Latin, but occasionally with Norse and Germanic influences. Monday is simply Moon's day, although in other languages the name (*lundi*, *lunes*) is derived from the Latin Luna. Tuesday means "Tiw's day" from the one-handed Norse god Tiw, roughly equivalent to the Roman god of war Mars. Wednesday is Woden's day: Woden (or Odin) was a Germanic god associated, like Mercury, with poetry and music. The Germans ignore Woden and call Wednesday *Mittwoch*, meaning mid-week.

Thursday is Thor's day, where Thor, like Jove, was the god of thunder; the Germans call it *Donnerstag*, which means "thunder day". Friday is Frige's day, where Frige (or Frigga) was an Anglo-Saxon goddess, Woden's wife, and equivalent to Venus, the Roman goddess of love, beauty, and sex. Saturday is Saturn's day, which in Spanish became *sábado*, the day of the Sabbath, and Sunday is the Sun's day, which in many languages has become "the Lord's day" – hence *dimanche*, *domingo*, and so on.

	GOD	ROMAN GOD	FRENCH	SPANISH
Monday	Moon	Luna	lundi	lunes
Tuesday	Tiw	Mars	mardi	martes
Wednesday	Woden	Mercury	mercredi	miércoles
Thursday	Thor	Jove	jeudi	jueves
Friday	Frige	Venus	vendredi	viernes
Saturday	Saturn	Saturn	samedi	sábado
Sunday	Sun	Sol	dimanche	domingo

The Gregorian calendar eventually put most of the world on the same footing as far as the date was concerned, but time was rather slower to come to heel. In the days before rapid communications, no one needed to know what the time was in the next village, let alone the next country. Everyone lived by local time – working during daylight hours, and taking noon to be when the Sun was highest in the sky.

STANDARDIZING TIME

SAILORS AND OTHERS WHO LIVED BY THE SEA cared about time, but what mattered most to them were the tides – low tide for foraging on the shore, high tide for setting sail. They needed a good knowledge of when the next high and low tides would be, but even that was linked more to the Sun than to any official time.

For professional mariners, and especially the British navy, regular time signals were introduced in the early 19th century. Tide tables were being produced, and so ships' captains needed to know the time in order to work out when to sail or to dock. More important, they needed to be able to set their chronometers to the correct time before they set sail, and the best place to do that was the Royal Greenwich Observatory.

King Charles II of England founded the Royal Society soon after he came to the throne in 1660, and was enthusiastic about science. He was persuaded that his navy would be stronger if the sailors had better navigational aids, and especially

BELOW: The need to know when to sail on the evening tide was one factor in driving the standardization of time.

more information about the stars. So in 1675 he commanded that an observatory be built on a hill overlooking the River Thames, and appointed John Flamsteed the first Astronomer Royal, who was to "apply himself with the most exact care and diligence to the rectifying of the tables of the motions of the heavens, and the places of the fixed stars, so as to find out the so much desired longitude of places for the perfecting of the art of navigation."

The observatory was designed by Christopher Wren, and paid for by a sale of £500-worth of old gunpowder, which according to legend was left over from the Gunpowder Plot of 1605. When finished (£20 over budget) the observatory was the first purpose-built research centre in Britain. Today it is mainly a museum and tourist attraction.

Two accurate clocks were installed in the new observatory, and Greenwich became the source of accurate time, especially for London. Messengers would come up the hill, set their own clocks or watches by the Greenwich clocks, and then carry the time back to their bosses in the City.

ABOVE: The Royal Observatory at Greenwich displays its famous time ball on the tower above the Octagon Room.

Greenwich Mean Time

Greenwich Mean Time (GMT) began with the observatory in 1675. For the next 200 years of international maritime trade there were many rival time centres, including Paris, New York, and several in Spain and Russia, but Greenwich was used more and more widely until it was officially adopted as the zero of longitude at the International Meridian Conference in 1884, and GMT became the standard time for the world (except for the French, who hung on for a few years). Ships' captains would keep a chronometer set to Greenwich time to calculate their longitude, from the difference between GMT and local solar time.

THE GREENWICH TIME LADY

In 1836 John Belville began a time service. He worked at the Greenwich Observatory, and so was able to set his pocket watch accurately every morning. The watch was a pocket chronometer which he called Arnold, because it had been made by John Arnold. In the afternoon he set off in his buggy and sold the precise time to his City clients – mostly clockmakers – who eventually numbered 200. When John died in 1856, his widow Maria took over the business, and when she retired in 1892 their daughter Ruth carried it on. Ruth used to go to the observatory at 9am every day, and get Arnold certified as being accurate to within a tenth of a second; then she went back down to the City and showed Arnold to all her clients. She became known as the Greenwich Time Lady, and she went on selling time until 1940, when she was 86 years old.

OPPOSITE: The clock on the gate of the Greenwich Royal Observatory shows Greenwich Mean Time on a 24-hour dial.

Greenwich time is mean because Earth is *tilted* and lurches nearer to and **further** from the **Sun**

BELOW: A modern time ball can be seen in Times Square, New York City.

Time balls

For the benefit of the captains of the ships moored in the Port of London, the Astronomer Royal John Pond installed a time ball on the observatory roof in 1833. This is a red metal ball, about 1m (3ft) in diameter, which every day slides halfway up a pole at about 12.55pm, to the top at 12.58pm, and then drops at exactly 1.00pm (GMT in winter, BST in summer), so that anyone within sight can set their clocks and watches precisely.

At the US Naval Observatory outside Washington DC, beside the white telescope dome, there is a golden ball which drops at noon every day – and has done so since 1845 – to provide a time signal for the captains of ships on the Potomac River. There are also time balls in 60 other places around the world, including Fremantle (Western Australia), Gdansk (Poland), Lyttelton (New Zealand), New York City, Quebec City (Canada), and Sydney (Australia). Today all these time balls operate mainly for show, since accurate time signals are transmitted by radio.

Railway time

People who were not sailors cared much less about precise time-keeping: they generally lived by the Sun, and if they needed to know the time they would probably look at the church clock, which was set so that it struck 12

WHY GREENWICH TIME IS "MEAN"

"True solar time" is the time as measured by the position of the Sun, and in particular 12 noon is defined as the time when the Sun is at its highest point in the sky. From 12 noon on one day to 12 noon on the next is 24 hours. There is a slight problem, however: this period is not constant from month to month, for two reasons. First, the Earth's orbit is not a circle but an ellipse, which means that the Earth is closer to the Sun early in January (146.2 million km; 91.4 million miles) than in July (151.2 million km; 94.5 million miles). When closer to the Sun, the Earth moves more quickly through space, and so the days are shorter.

Second, the axis of the Earth is tilted, so that the Sun appears to move in a circle that is not parallel to the equator except at the equinoxes. The Sun effectively moves faster at the equinoxes, which means that the days are shorter in March and September than in June. Greenwich Mean Time is an average of solar time taken over the year, as measured at Greenwich. It differs from true solar time, as measured on a sundial, by as much as 16 minutes a day fast on 4 November and almost 15 minutes slow on 12 February. On the other hand, it agrees with true solar time on 15 April, 13 June, 1 September, and 25 December. This variation is called the "equation of time" and has been known and understood for centuries.

SHEPHERD PATENTEE

53 LE____L ST. LONDON

GALVANO-MAGNETIC CLOCK

THE SHEPHERD 24-HOUR GATE CLOCK

This is one of the earliest electrically-driven public clocks and was installed here in 1852. The dial always shows Greenwich Mean Time (GMT) or Universal Time (UT). In summer Britain converts to British Summer Time (BST), which is an hour ahead of GMT, and the clock then appears one hour 'slow'.

Being a 24-hour clock, the hour hand marks noon (XII) at the bottom of the dial and midnight (0) at the top. The time shown is accurate to 0.5 of a second.

THE TIME BALL

The red Time Ball on top of Flamsteed House is one of the world's first visual time signals. It was installed in 1833, though the present one dates to 1919, as a signal for ships in the Thames to check their marine chronometers.

The Time Ball drops daily at 1300hrs (GMT in winter, BST in summer). It is raised halfway up the mast at 1255hrs as a preparatory signal and to the top 2 minutes before it drops.

HEIGHT ABOVE
MEAN SEA LEVEL
AT NEWLYN
154·70 FEET.

ORDNANCE SURVEY BENCH MARK

The small plate (marked 'G 1692') is an official bench mark, used by surveyors for fixing up their levels and fixed here in 1957.

The top of the signal marks 152ft above Mean Sea Level at Newlyn, Cornwall, which is the national Mean Sea Level Bench Mark set up by the Ordnance Survey in 1921-2.

PUBLIC STANDARDS OF LENGTH

These British Imperial standards were first mounted outside the Observatory main gates some time before 1866.

To test the accuracy of a ruler, place it on the appropriate pair of metal pins. The correct length is the distance between the inner faces of the D-shaped studs.

THE MECHANICAL TELEGRAPH

The time ball was useful only to those who could see it, but in 1796 the churchman George Murray had built for the British Royal Navy an efficient mechanical telegraph system to send messages to other ports, especially Portsmouth, home of the country's warships. There were stations 15–30km (10–20 miles) apart positioned so that there was a line of sight from one to the next, and connecting the Admiralty in London with the naval base at Portsmouth. At each station was a big frame with six panels or shutters, each of which could be open or shut. There was a simple alphabetical code, so that A was "all shut except top left", B was "all shut except top right", and so on. Men were employed to operate the shutters with ropes, and as soon as they saw a signal from the London direction they repeated it for the benefit of the next station. A complex message might take many minutes to send, but every morning they sent a time signal, and were able to transmit it over the 112km (70 miles) from Greenwich to Portsmouth in less than one minute. Later, with the coming of the electric telegraph, these mechanical systems became obsolete.

BELOW: Railways brought the introduction of standard time, first in Britain, and by 1883, in North America.

o'clock when the Sun was at its highest in the sky. This casual attitude was utterly changed by the railways. The first train pulled by a steam locomotive ran in the Taff Valley in South Wales on 21 February 1804. To settle a bet, the fiery Cornish engineer Richard Trevithick hauled 10 tonnes of pig iron 16km (10 miles) from the Penydarren iron works to the wharf at Abercynon. The first dedicated passenger railway ran from Liverpool to Manchester in 1830, pulled by *Rocket*, which was designed by Robert Stephenson and driven by his father George, and within 10 years railway fever had taken hold, and railway lines were snaking all over England.

In 1841, Isambard Kingdom Brunel's Great Western Railway (GWR, sometimes called "God's Wonderful Railway") reached Bristol, and that was when the trouble started. People going to Bristol Temple Meads station hoping to travel to Swindon or London kept missing the trains, which seemed to leave 11 minutes early.

The problem was that the engine drivers had come from London, and were using London time. Bristol is 320km (200 miles) west of London, which means that the Bristol sunrise is 11 minutes later than the London sunrise and Bristol time was therefore 11 minutes behind London time. The only sensible solution was for all railways to keep London time, or "railway time", although many places kept their own time as well. In Bristol, some public clocks had two minute hands, one for local time and one for railway time, set 11 minutes apart.

Railway time was the first sort of time that was standardized for ordinary people. In 1847 the English railways adopted Greenwich Mean Time as their standard, and by 1855, clocks all over England were standardized to GMT.

Railway time followed a precedent set some decades earlier by the General Post Office, known as GPO time. Mail coaches were always accompanied by a Mail Guard, who carried a Time Bill, which specified stopping points and times of the coach, and a locked Time Piece set to London time. The Post Office did not officially switch to GMT until 1872.

This standardizing revolution came a bit later in North America. The US Navy established a depot which in 1854 became the US Naval Observatory and Hydrographic Office (USNO) and is now the home of US time. The problem of standard time was a great deal more difficult in North America than in England. In England, every town could move to London time, but in the US it was clearly absurd for every city to adopt, for example, New York time, when the western states were several hours behind. During the winter it might be noon in New York, but the people of Seattle and Vancouver would still be waiting for sunrise. The Canadian railroad engineer Sir Sandford Fleming invented the idea of time zones, and proposed to use them to standardize the railroad schedules across the country. The heads of the major railroads met in Chicago and agreed to adopt the Standard Time System of four time zones for the continental USA. On 18 November 1883, standard time was imposed across the United States and Canada.

ABOVE: In the 1840s, clocks in many English provincial train stations had two minute hands: one for local time and one for railway time.

In the 1840s Great Britain was the first country in the world to *run on railway time*

American physicist Richard Feynman imagined the Earth, as seen from outer space, spinning on its axis in the sunshine, with a line curving round the globe dividing the night from the day. All along this line, people are getting up and brushing their teeth. As the Earth turns, that line of tooth-brushers moves steadily westwards, spinning with the Earth.

GOING GLOBAL

ABOVE: Where in the world does the day begin and end?

EVERY HOUR, THE LINE WILL MOVE 15 degrees to the west, and starting in New York and Toronto will reach Houston and Saskatchewan after one hour, Colorado and Alberta after two hours, California and Vancouver after three, and Alaska after four.

Before 1883, each US city used its own local time, but from that date the time was fixed in blocks across the nation, although the idea took hold rather gradually. In Detroit, for example, the City Council decreed in 1900 that the clocks should be put back 28 minutes to Central Standard Time, but half the city refused, and

RIGHT: The world's time zones. The time changes suddenly every 15 degrees of longitude west or east.

eventually the whole city went back to local Sun time. There was even an offer to put up a sundial in front of city hall. Standard time was not established by federal law until 1918. Today there are five time zones across mainland USA: Alaska time (4 hours behind Eastern time), Pacific (3), Mountain (2), Central (1), and Eastern.

Around the world there are 24 principal time zones, which are more or less bounded by lines of longitude, although the boundaries are often bent to take in whole states or countries. Time zones were based on Greenwich Mean Time (GMT), and Greenwich in England is the prime

meridian, the zero of longitude. Every 15 degrees of longitude to the west, the time zone changes by 1 hour, so that 360 degrees corresponds to 24 hours. Each time zone covers a wide stretch of territory, so that the official time accurately matches local time only in a narrow band, but for convenience the entire zone uses the same time, and there is a sudden change at each boundary.

At 180 degrees longitude, or thereabouts, is the International Date Line (IDL), where not only the hour but the date changes abruptly. If you were to fly eastwards from the eastern tip of Russia to the western tip of Alaska, the time would jump from 9am Tuesday to 9am Monday, and you would have the rest of Monday and Monday night all over again. Flying the other way, the time would jump from 9am Monday to 9am Tuesday, and you would lose a whole day. I once flew from Tokyo, Japan, to Hawaii. We took off on Sunday morning, flew all day over the Pacific Ocean, and arrived in Honolulu on Saturday evening. Most strange. Having two places a few miles apart with differing days would be rather confusing, so the IDL has been deliberately kinked to avoid going through land.

Because Greenwich was the accepted zero of longitude, and Zulu is the name for the letter Z in the NATO phonetic alphabet, GMT came to be called "Zulu time" by radio operators. They would say "1545 Zulu", meaning 3.45pm GMT, and this could be written 1545Z. This remains the case for pilots, who use Zulu time wherever they are flying, to avoid misunderstandings caused by time zones.

BELOW: The Kiribati Islands in the Pacific unilaterally moved the date line, so that all the islands could enjoy the same day together.

International date line

Pacific Ocean

The Kiribati Islands

Australia

The Greenwich time signal

On 5 February 1924, the British Broadcasting Corporation (BBC) first transmitted a time signal by radio, which allowed anyone who could receive the signal to set their clocks, watches, and chronometers. Radio waves travel at the speed of light, so even if the signal had to go halfway round the world it would still arrive within a tenth of a second, which is good enough for anyone setting a watch or clock by hand.

The time signal takes the form of six "pips" or short tones (of 1kHz) at one-second intervals; the last pip is slightly longer than the others. Technically, each pip begins at the exact second (eg 6.59.55, 6.59.56, etc) and lasts for one tenth of a second. The sixth pip starts at the exact hour (7.00.00 precisely) and lasts for half a second. When there is a leap second, it is indicated by a seventh pip.

This signal is still broadcast by the BBC on many of its channels, and by many other broadcasters around the world, although the advent of digital radio presents a problem, because the signal is delayed by a second or two, and therefore not much use for precise time-setting.

The pips were originally broadcast from Greenwich, and are still called the Greenwich Time Signal, although now they are generated by the National Physical Laboratory at Teddington and are governed by UTC (see opposite). The US Naval Observatory transmitted time signals by electric telegraph from 1865, and the first radio time signals in 1904.

RIGHT: Built in 1970, the Greenwich Time Service atomic clock relayed its signals to the BBC so that they could broadcast their famous "pips".

BELOW: An electronic display shows the "pips". Each one starts at an exact second, and the longer sixth pip begins exactly on the hour.

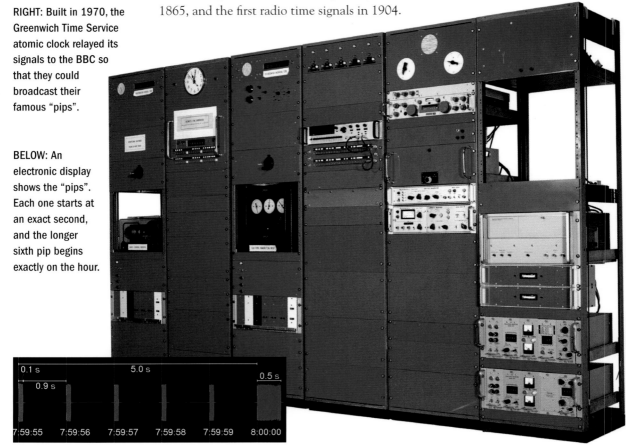

LEFT: The master clock of the family of atomic clocks is held at the US Naval Observatory outside Washington DC.

Coordinated Universal Time was invented to cope with the Earth's w o b b l i n g

Coordinated Universal Time (UTC)

GMT was good enough for most of the world until the second half of the 20th century, but then astronomers suggested that there should be a more stable standard. The problem is that the Earth does not spin smoothly: it wobbles a bit, like a fat person on a dance floor. What is more, its spin is gradually slowing down: because of the frictional drag of the tides sloshing to and fro, the Earth takes about 2.3m (7.5ft) per second a day longer to turn every hundred years.

On the other hand, the melting of the continental ice sheets at the end of the last ice age caused a "glacial rebound" of the land underneath, and this has the effect of speeding up the Earth's rotation by 0.6ms/day/century. The net rate of slowing is therefore 1.7ms/day/century, which is the average observed for the last 2700 years.

On 1 Jan 1972, GMT was officially replaced by Coordinated Universal Time, which is maintained by a collection of 260 atomic clocks in 49 places around the world. For most people GMT and UTC are the same time, and many organizations still refer to GMT. UTC is kept to within 1 second of GMT by the addition of the occasional leap second. For example, 31 December 2008 became 86,401 seconds long rather than 86,400 seconds.

WHY IS IT CALLED UTC?

The International Telecommunication Union wanted Coordinated Universal Time to be accepted by everyone and therefore in all languages. English-speakers wanted CUT; French speakers wanted TUC (Temps Universel Coordonné); so they compromised on UTC.

BENJAMIN FRANKLIN

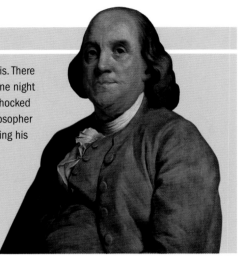

In 1784, Founding Father of the USA Benjamin Franklin was 78 and living in Paris. There he wrote a witty article called "An Economical Project" for the *Journal de Paris*. One night he had not gone to bed until 4am, but was woken by a noise at 6am, and was shocked to find the Sun was shining and streaming into his room. He claimed that a philosopher had assured him there could be no light at that hour of the morning, so by opening his shutters he could not have let the light in: he must have let the darkness out.

   Nevertheless, he calculated that if everyone in Paris were to get up at sunrise during the summer, and use sunlight in the morning instead of candlelight in the evening, they would save more than 30,000 tons of candles every year. He proposed that candles should be rationed, shutters should be taxed, and people should be woken at dawn by bells and cannons.

BELOW: The retailers United Cigar Stores celebrate the 1918 US daylight saving bill.

Daylight saving

The idea of daylight saving was first mentioned in an essay by the American scientist and diplomat Benjamin Franklin, who coined the phrase "Early to bed and early to rise makes a man healthy, wealthy, and wise." His proposal was made in jest, and was not taken seriously; nor were the serious suggestions made by New Zealand bug-hunter George Vernon Hudson in 1895, and by the builder and golfer William Willett and politician Robert Pearce in England in 1907–9. In 1916, however, daylight saving was put into effect in Germany and Austria, and spread rapidly across Europe, including Great Britain, where it is called British Summer Time (BST).

   Canadian provinces Manitoba, Newfoundland, and Nova Scotia soon followed, and daylight saving was introduced in the USA along with the Standard Time Act of 1918. The idea is to allow people to enjoy more of the summer evenings before bedtime, by moving an hour of daylight from the early morning to the evening, when it is more useful, or in other words shifting clock time one hour forward relative to the Sun.

   Technically what happens is that in the USA the clocks go from 1.59am to 3am on the appointed day in the spring, and from 1.59am to 1am in autumn. In Europe the change happens at 1am rather than 2am.

   Daylight saving time (DST) generally begins early in March and ends late in October, but the whole idea is contentious, and not only do the dates keep shifting, but at first some places refused to use daylight saving at all. This often resulted in chaos: on a 56km (35-mile) bus journey from Moundsville, West Virginia, to Steubenville, Ohio, the drivers had to reset their watches seven times. In 1965, the twin cities of Minneapolis and St Paul chose different dates to start DST, which caused chaos for everyone who was doing anything on both sides of the Mississippi River. Now DST is mandatory in all states apart from Arizona and Hawaii.

Daylight saving time
was first suggested as a joke by Benjamin Franklin in 1784

Twins can suffer from DST. In North Carolina in November 2007, a woman called Laura gave birth to a boy at 1.32am, and then to his twin sister 34 minutes later. But DST had started between the two births, so on paper the girl was born at 1.06am. Let's hope the lawyers never have to rule on who was born first: after say 60 years, with memories fading and no surviving witnesses, it might be a tricky decision.

There are perennial arguments about whether DST is of benefit to anyone. Proponents claim that they enjoy the long summer evenings, because DST shifts an hour of daylight from the early morning (when most people are asleep) to the evening (when they are awake). As a result, people have to switch lights on for fewer hours, and so may save energy. In 1986, DST was moved in the USA from the last Sunday in April to the first, and the Department of Transportation claimed that had saved 300,000 barrels of oil. In New Zealand, power consumption decreases by 3.5 per cent when DST begins.

BELOW: Minneapolis is across the Mississippi River from St Paul. In 1965, the two cities caused chaos by starting daylight saving time on different days.

ABOVE: DST allowed increased productivity in wartime by taking advantage of extra evening light in summer. In Britain during World War II, even greater productivity was achieved by "double British summertime".

When Indiana switched to DST, however, there was a significant increase in power consumption over the next three years. This may have been the result of the increased use of air-conditioners in those long, hot summer evenings. But most important is the fact that there is a small but significant (1 per cent) reduction in road traffic accidents during DST, although pedestrians turn out to be at more risk just after DST ends than while it is in force.

Opponents of DST say they hate having to get up earlier, and if it starts too early in the year, children have to go to school in the dark, which is dangerous. Furthermore, there are now clocks embedded in most of the gadgets in the house, so when DST begins you have to change not just the clock on the mantelpiece but also the ones in the cooker, the microwave, the radio, the TV, and the central heating, as well as your wrist watch, your mobile phone, and the alarm clock by the bed. In addition, there is the recurring problem that some people forget that DST has started, and so turn up for appointments or meetings an hour late. Other people switch their clocks the wrong way… But there is an easy mnemonic to help you to move your clocks in the right direction – Spring forward; fall back.

The US Department of Transportation claimed that in a month DST saved **300,000 barrels of oil**

DST is now used across most of North America and some of South America, across all of Europe, Russia, and some of the Middle East. It has been tried and abandoned in most of South America, Australia, and the Far East, and it has never been tried in most of central Africa, for the simple reason that near the equator there are about 12 hours of daylight all through the year, so there is no point in instituting DST.

From the marking of midwinter by Stone Age people around the world to the present system of a standardized calendar and universal time that is precise to less than a second in a year, we have come a long way. How we got to where we are today was a result partly of human intelligence and ingenuity, and partly of the increasingly clever technology we have built over the centuries to record and measure the passage of that elusive stuff, time.

People, and most other animals and plants, have always lived by the day, getting up with the sun and sleeping during the night. For millions of years no one bothered to measure the day – it measured itself and set the rhythm of life.

DAYS

A DAY HAS ALWAYS BEEN the time between consecutive noons, when the Sun is exactly overhead. Measuring the time between consecutive sunsets (or sunrises) is easier, but that interval is not constant; outside the tropics it increases during the autumn and decreases during the spring.

ABOVE: After spending three years underwater as nymphs, mayflies emerge into the sky to breed. They spend just one single day in their adult (imago) stage before they die.

How long is a day?

Even the time between successive noons is not constant, because the Earth's orbit is elliptical rather than circular. As a result the length of the day varies through the year by up to a minute; the mean Earth day is the average length of a day, or 86,400 seconds. Technically this is called a solar day or a tropical day; solar because it is based on the time between successive moments when the Sun is overhead, and tropical because as the seasons pass the Sun appears to turn round each time it reaches the tropics; the Greek for turn is *tropikos*. The stellar or sidereal day is used by astronomers, because it is the time interval between the consecutive appearances of a particular star overhead. This is about 4 minutes less that the solar day, or to be precise 23hrs 56mins 4.1s.

The days used to be shorter; they lasted only 22 hours, and there were 400 days in the year only 620 million years ago.

When is a day longer than a year?

Day length varies from planet to planet: Mercury 59d, Venus 243d, Mars 25hrs, Jupiter 10hrs, Saturn 10hrs, Uranus 18hrs, Neptune 19hrs. Curiously, all the enormous gas giants spin much faster than little rocky Earth; Mars spins at almost the same rate as we do; only Mercury and our sister planet Venus spin much more slowly. On Venus the day is much longer than the year.

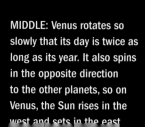

MIDDLE: Venus rotates so slowly that its day is twice as long as its year. It also spins in the opposite direction to the other planets, so on Venus, the Sun rises in the west and sets in the east.

620 million years ago there were

400 days in one Earth year

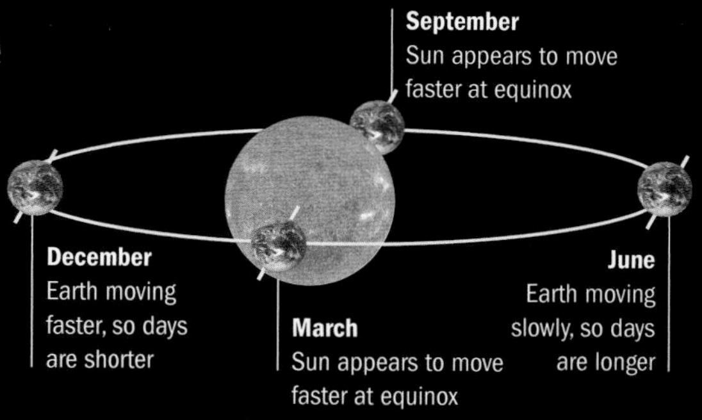

September
Sun appears to move
faster at equinox

December
Earth moving
faster, so days
are shorter

March
Sun appears to move
faster at equinox

June
Earth moving
slowly, so days
are longer

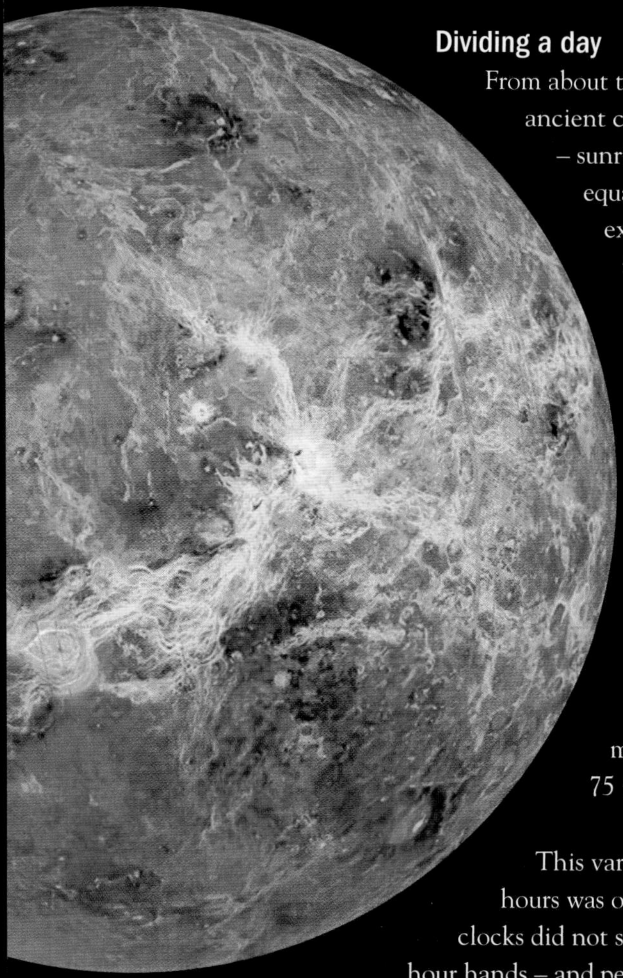

Dividing a day

From about the 2nd century BC, many ancient civilizations divided the daylight – sunrise to sunset – into twelve equal hours. The Romans, for example, numbered their hours from sunrise, so that *hora prima* was the first hour of daylight, and was followed by *hora secunda*, *hora tertia*, and so on, with the seventh hour *hora septima* starting at noon. This meant that the hours varied considerably in length, both from month to month and from place to place. For Julius Caesar, in Rome, the hours would have been approximately 45 of our minutes long in midwinter and 75 in midsummer.

This variation in the length of the hours was one reason why ancient clocks did not show minutes; they had only hour hands – and perhaps life was less frantic.

ABOVE: The length of the day varies through the year as a result of the eccentricity of Earth's orbit – it is an ellipse rather than a circle. The days are shortest in December and longest in June.

IMAGINING A DAY

A Venus sidereal day is 243 Earth days long. However, the Sun rises and sets twice in that time. A solar day on Venus lasts

Venus				
Midnight	6am	Noon	6pm	Midnight

Earth				
January 1	January 30	February 28	March 30	April 27

CHAPTER 4

MEASURING TIME

Accuracy and precision are different concepts when applied to clocks. Suppose that a clock keeps perfect time, counting exactly 31,556,900 seconds in a mean solar year, but remains a permanent 20 minutes slow when compared to Greenwich Mean Time. This clock is extremely precise, but not accurate. The concept of accurate time is anyway somewhat philosophical, since it implies agreement with a standard – but there are several different time standards, all of them arbitrary. No clock can be perfectly accurate with respect to all of them. In this chapter, I use "accurate" only when I am comparing a timekeeper with a particular standard

The Sun is the most obvious marker of the passage of time, and people who wanted to divide the day into regular periods must long ago have started using shadows.

SOLAR TIME

A STICK THRUST INTO THE GROUND will cast a shadow whenever the Sun is shining, and the direction and length of the shadow carry useful information about the progress of the day. The shadows are longest at sunrise and sunset, and shortest at midday: placing pebbles or other markers at the end of the shadow at intervals through the day will quickly reveal the time of midday, and can be used to divide the day into shorter intervals, such as hours. The obelisks of ancient Egypt may well have been erected partly to throw such shadows.

If you try using such a device you will soon discover that the shadow positions vary through the year. To get a consistent set of shadows you need to angle your stick, or "gnomon", so that it is parallel with the Earth's axis. This means that the angle between your gnomon and the horizontal should be the same as your latitude. In other words, a gnomon at either pole should be vertical, but at the equator, horizontal. In the northern hemisphere, your gnomon should point at the pole star Polaris. The shadow of the gnomon then moves around its base at a regular rate through the year, and can indicate hours that are engraved on it.

The oldest known sundial was built in Egypt around 1500 BC. In his 1st-century BC work *De Architectura*, the Roman engineer Vitruvius describes mathematically, using complex geometry, how to construct a sundial, and mentions a dozen different kinds.

THE SAMRAT YANTRA

One of the biggest sundials in the world, the Samrat Yantra, was built at Jaipur in India around 1730. The gnomon is 27m (89ft) high, and is angled at 27 degrees from the horizontal, which is the latitude of Jaipur. The Sun's shadow from this monster moves across the scale at about 1mm (1/25in) per second, or a hand's breadth in a minute, which is an impressive sight. This sundial was probably designed to deliver great precision: in theory it should be possible to read the time to within a couple of seconds. Unfortunately, however, the Sun is not a point source of light, and its shadow is therefore not sharp, but consists of a dark umbra in the centre surrounded by a soft-edged penumbra. This means that the exact moment when the shadow reaches the marks engraved on the bronze scale is almost impossible to gauge.

Sundials are commonly made with horizontal bases, but some are built vertically on the walls of churches or other old buildings. Some use a line of light to indicate the time, rather than a shadow; others have mirrors or lenses. Some take the form of hollow spheres, with the shadow from an axial rod falling on the inside of an equatorial band; others are in the form of cylinders or cones. All, however, rely on the rays of sunlight to provide an indication of the time.

In the northern hemisphere, a sundial's gnomon should point at the
pole st★r

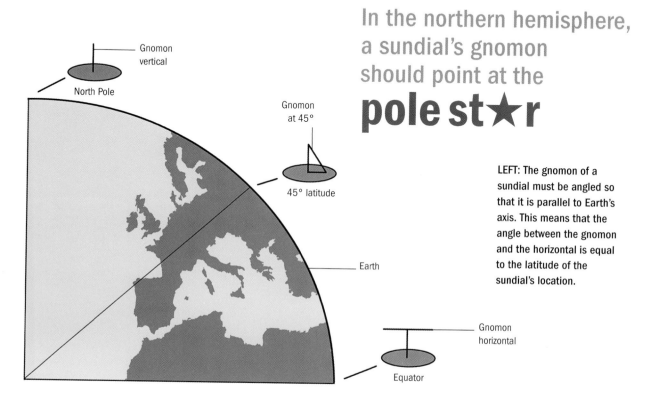

Gnomon vertical

North Pole

Gnomon at 45°

45° latitude

Earth

Gnomon horizontal

Equator

LEFT: The gnomon of a sundial must be angled so that it is parallel to Earth's axis. This means that the angle between the gnomon and the horizontal is equal to the latitude of the sundial's location.

Sundials are regular and reliable, but they are useless in cloudy weather and at night. Telling the time by the stars is possible at night if the sky is clear of clouds – but this needs knowledge and experience. Different types of time-keeper were needed to "carry" the time through the night and through cloudy weather, and some of the earliest were water clocks.

WATER CLOCKS

ABOVE: A simple ghati, made from a coconut shell, with a hole to allow the water to flow in.

BOTTOM RIGHT: An ancient Greek klepsydra allowed water to trickle out through a hole in its base.

IN ANCIENT INDIA THE STANDARD WATER CLOCK was the ghati. Similar timing devices were also used in many parts of Africa. To make a simple ghati, drill a small hole in the middle of a half coconut shell, and float it in a bowl of water. Gradually the water will leak in through the hole, and in due course the shell will sink. In an Indian temple, the monk would no doubt have had a finely engraved copper ghati, perhaps 20cm (8in) in diameter, rather than a half coconut shell, and the hole would have been drilled to just the right size so that the ghati would sink in 24 minutes, which was the length of the Indian hour. The day comprised 60 hours of 24 minutes, so the minutes were the same length as they are today.

Traditionally the monk would set the ghati floating, and sit down in meditation. At the moment when it sank he would strike a gong or a drum, then refloat the ghati, and meditate once more. According to legend the monks became so used to the procedure that they could sound the gong at 24-minute intervals without even using the ghati. Presumably in practice they had two ghatis, so that they could float the second the instant the first one sank, and then take a few moments to retrieve the first one.

Water clocks were also used in Babylon and in Egypt 1800 years ago, and continued to be used throughout the eastern Mediterranean for many hundreds of years, but these clocks were in a sense the opposite of the ghati: instead of measuring how long it took for water to fill a container, they measured how long it took to run out. The Greeks called such a clock a klepsydra (water thief).

Water thief

The simple klepsydra was a jar with a small hole near the bottom. The jar was filled with water, and a lawyer or defendant in court was allowed to speak until the water had all run out. This provided a simple fixed measure of time, and was therefore useful not only for court

The ghati,
a traditional Indian water clock, sinks in 24 minutes

BELOW: The Elephant Clock described by al-Jazari in the 13th century was a design for a giant weight-powered water clock riding on an elephant.

appearances but also for prostitutes who wished to have a regular time to entertain their clients – according to legend one well-known lady of the night called herself Klepsydra.

Any particular klepsydra could measure only one time interval; no adjustments were possible. This caused Aristotle famously to remark that the length of a play should be governed by the plot, and not by the klepsydra. The klepsydra was poor for measuring the passage of time, because as the water level in the jar dropped so did the pressure at the bottom, and the water ran out increasingly slowly. One way of getting round this was to make the jar conical, with a wide top and a narrow bottom, so that when the water was high, a greater amount had to flow out in order to drop the same distance as when the water was lower in the cone. Klepsydras of this design were still being made by the great Muslim engineer al-Jazari around AD 1200. He wrote a *Book of Knowledge of Ingenious Mechanical Devices* in 1206. There he described several grand clocks, including the Elephant Clock and the Castle Clock, which have recently been reconstructed. During the intervening centuries, however, the klepsydra had been developed and improved in other ways as well.

In the 3rd century BC a Greek called Ktesibios, who lived in Alexandria, devised a way to avoid the problem of

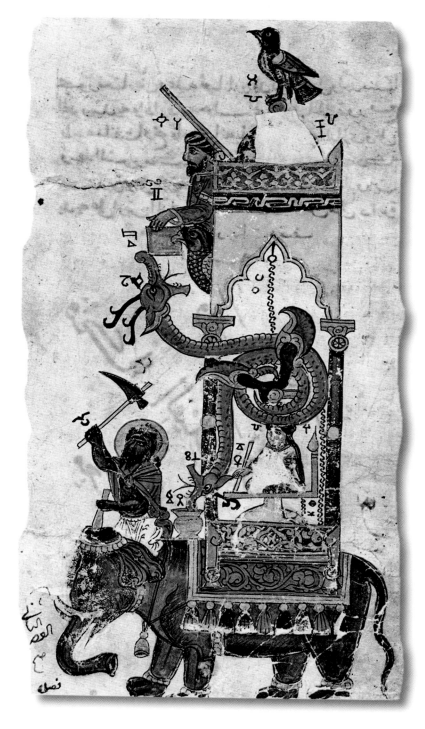

KTESIBIOS

Ktesibios (flourished 285–222 BC) was a born tinkerer and inventor. His father was a barber who used a mirror to show his clients how smart their haircut was. The mirror was heavy, and Ktesibios arranged that it should hang from a rope which passed over a pulley and was supported by a counterweight, so that the mirror was easy to manoeuvre. He put the counterweight inside a tube, to prevent its swinging about dangerously, and noticed that it made a hooting note as it slid up and down and drove the air out of the tube. Ktesibios developed this into a water organ, or hydraulis (seen right, along with a trumpet), which was the world's first keyboard instrument, and remained a favourite of Roman emperors for centuries, as it was loud enough to be heard in the arena at games and other events.

the pressure dropping in the klepsydra. He arranged that the jar was always full of water, by having a continuous supply, and an overflow, rather like a modern garden water-butt. The flow from the bottom of the jar was then regular, and he arranged for the water to fall into another container, where it gradually raised a float. A pointer on the float marked the hours, so the "improved klepsydra" could better measure the passage of time. Once he had this rising float – a steady mechanical movement – Ktesibios was able to make it do all sorts of things, including sound alarms, blow whistles, and set off mechanical contraptions.

His clocks were elaborate and precise, but they never showed minutes, nor any other divisions of the hour. This was partly because time was less precious in those days, so that no one cared about minutes, and partly because the Greeks, like many other peoples, divided the day into 12 equal hours, which were therefore longer in summer than in winter. The scale could be varied to allow for hours of various lengths, but tracking minutes was deemed unnecessary.

Harun al-Rashid

Water clocks were the best available timepieces for several hundred years, and many elaborate ones were built, some as gifts. Harun al-Rashid, the fifth and most famous Abbasid Caliph, and star of *The Thousand and One Nights*, ruled from AD 786 to AD 809. He lived first in Baghdad, where he founded the House of Wisdom, and then in Ar-Raqqah. He was a contemporary of Charlemagne, who was crowned Holy Roman Emperor on Christmas Day AD 800.

Harun al-Rashid sent Charlemagne presents, including silk, perfume, ivory chess pieces, an elephant, and a water clock that every hour dropped

BELOW: Harun al-Rashid is depicted in a scene from *The Thousand and One Nights*.

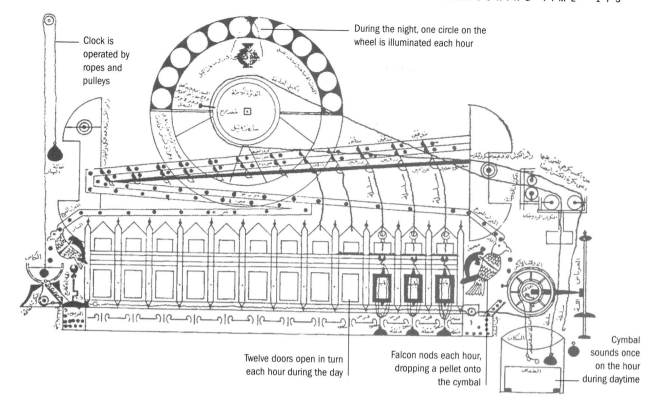

Clock is operated by ropes and pulleys

During the night, one circle on the wheel is illuminated each hour

Twelve doors open in turn each hour during the day

Falcon nods each hour, dropping a pellet onto the cymbal

Cymbal sounds once on the hour during daytime

a bronze ball into a bowl, while an armed figure came out from a door. Charlemagne was fascinated by this magical device, especially by the noises and tricks it displayed on the hour.

Near the Umayyad Mosque in Damascus, the Jayrun water clock was built. It was described by Ridhwan al-Sa'ati in AD 1202, although he was only its operator: it had originally been built at least 50 years earlier. It was based on the principle of the ghati, with water leaking into an empty container.

This clock divided the day into 12 equal hours, throughout the year. On each hour one of the 12 brass doors opened to reveal which hour had passed, and two falcons threw copper balls into a large copper cup to ring out the hour. Below each door was an indicator to show how much of the hour had passed, with divisions five minutes wide. The clock also indicated the signs of the zodiac, the angle of the Sun during the day, and the time at night, with lights.

Zhang Heng

Meanwhile, the ancient Chinese had also invented water clocks, using principles similar to those of the klepsydra. During the 1st century AD, the brilliant inventor Zhang Heng followed Ktesibios in arranging a constant flow into the receiving vessel. Six hundred years later, Yi Xing produced an escapement mechanism (see "Clock Fundamentals" p180), which was then used by Su Song in 1088 to drive an amazing astronomical clock in a 10m (33ft) tower. This machine had a water-powered bronze armillary sphere, a rotating celestial globe, and five panels in front which opened to show figures holding plaques to declare the time of day.

ABOVE: The Jayrun water clock was described in the treatise *On the Construction of Clocks and their Use*, written by Ridhwan al-Sa'ati in the 13th century. The clock was powered by falling water, whose motion was transmitted to various moving parts by a system of ropes and pulleys. Each hour of the day or night was marked by a change on one of the clock's two dials – one for daytime and the other for night-time.

About 2000 years ago, the Romans used candle clocks, of which the simplest type were candles with Roman numerals inscribed on them. When the wax burned down to VII it was seven o'clock, or at least the beginning of the seventh hour.

Burning clocks and sand glasses

THE ROMANS DIVIDED THEIR DAYLIGHT into 12 hours of equal length, which were therefore longer in summer than in winter. This meant that they needed a variety of candles to cope with the varying hour lengths. The first hour after sunrise was called *hora prima* (first hour), and in midwinter the seventh hour of the day, *hora septima*, would have run (in our time) from about 12 noon to 12.45. Away from Rome in more northern or southern parts of the Roman Empire, a different set of candles was needed, adjusted to the changing day length at that latitude.

Candle clocks were used for hundreds of years in many countries, including China, Japan, and England, where King Alfred is recorded as having a candle clock around AD 878. These candles could easily be made into alarm clocks by pushing a nail into the wax at the wanted hour: when the hour came, the nail would fall out onto a gong or bell. Because draughts would make the candles burn unevenly and too fast, the candles were probably encased in boxes with transparent windows.

The Muslim engineer and polymath al-Jazari described a clever candle clock in which the candle was balanced by a counterweight. As the candle steadily burned away, the counterweight pulled it upwards, and operated automata which revealed the time of day.

Oil lamps were also made into clocks. Provided the oil burned at a steady rate – the flame would need protection from draughts – the rate of fall of the oil level in the reservoir would give a measure of the passage of time.

ABOVE: An 18th-century candle clock marks the hour as it burns down.

Sand glasses

Neither water clocks nor candle clocks were reliable in the wild, wet, and windy conditions on board ship. To measure their speed through the water, sailors used a "log line". This was a long piece of rope with a series of knots tied in it at regular intervals and a chunk of wood (a "chip log") on the end. They threw the wood overboard so that it was left behind in the sea and it began to pull the rope over the stern after itself. The sailors counted the number of knots that passed over the stern in a fixed time.

One standard system was to tie the knots every 14.4m (47ft 3in) along the rope and count the number of knots that passed in 30 seconds. If five knots had passed, then the speed of the ship was 5 knots, where 1 knot is about 0.5m/s, 1.85km/h, or 1.15mph. Knots are still used as units for the speed of ships and aircraft today.

To measure the 30 seconds, the sailors used a sand glass. This comprises two glass bulbs joined by a narrow neck, with some dry sand inside. When the sand glass is stood on one end, the sand takes a fixed time to run through to the other bulb. Sailors used a 30-second sand glass, but they can be made to show any fixed period of time – there are also "hour glasses" – and they are not much affected by dampness, draughts, or even the rolling of a ship.

Because they measure a fixed time interval, sand glasses are still used, rather like the klepsydra once was, for timing boiled eggs, for meetings, and for other events that should run for a fixed time.

BELOW: A chip log tied to the end of a log line was thrown overboard behind a ship. Sailors timed how fast the rope was pulled into the water to measure the ship's rate of knots.

Some early water clocks had a number of elaborate mechanisms built into them, but they were all powered, in one way or another, by the flow of water, and controlled by the rate of flow through a hole or holes. The next major step in clock design came with the use of falling weights, rather than water, as the source of power, and with mechanical oscillators for regulation.

MECHANICAL CLOCKS

THE EARLIEST MECHANICAL CLOCKS WERE BUILT in the 13th or 14th century. A few were for domestic use by rich homeowners, and hung high on the walls to allow space for the weights to fall, but most were massive beasts made of cast iron, often with huge rocks as weights to drive them.

In Christendom, the clocks were usually paid for by the Church, and placed in cathedrals or monasteries; their main function was to call people to prayer at specified times. The clocks had no hands, nor any other means of displaying the time, but rang bells on the hours. Indeed, the word clock comes from the Latin *clocca* meaning "a bell". The introduction of these great machines was probably due at least in part to the cloudy weather of Europe. Around the Mediterranean, sundials were sufficient, but under the clouds of Europe the faithful needed a more reliable stimulus to get them on their knees. Mind you, bells were probably more efficient for summoning people even in the Mediterranean climate.

LEFT: Strasbourg Cathedral's 1843 astronomical clock is the third clock on the site. Mechanized figures appear and ring bells when the hour is chimed.

In Strasbourg's Gothic cathedral, the Clock of the Three Kings, 18m (59ft) high, was built between 1352 and 1354. It had various devices, including an astrolabe and a calendar, but most spectacular was a gilded rooster, symbol of Christ's passion, which at noon flapped its wings and crowed, at which the three kings bowed before the figures of Mary and Jesus. The clock stopped working and was replaced in the 16th century by an even more elaborate machine. This also had a golden cockerel, which unfortunately was struck by lightning in 1640. The current clock was built by Jean-Baptiste Schwilgué in 1843.

Richard of Wallingford, Abbot of St Albans, built a clock for his abbey which was almost complete when he died in 1336. Its main purpose was to regulate the times of prayer and work for the monks, and it struck the hours, but it also had a primitive time display. As well as all this, it showed the phases and eclipses of the Moon, and the stars visible from St Albans at the time. Other notable medieval clocks include the one built around 1386 at Salisbury Cathedral, now the oldest working mechanical clock in the world; the Old Town Hall clock or Orloj in Prague (1410), which has a procession of apostles and a crowing cockerel; and the Wells Cathedral clock, completed in 1392. These clocks showed the positions of Sun, Moon, planets, and stars, and were known as astronomical clocks or astraria.

Giovanni de Dondi, who lived with his clockmaker father in Padua, completed an astrarium in 1364. It was about 1m (3ft) high, and used 107 gear wheels to show the positions of the Sun, Moon, and the five known planets, as well as the time on a 24-hour dial, and all the church feasts on a large calendar drum. It was described as being "full of artifice, worked on and perfected by your hands and carved with a skill never attained by the expert hand of any craftsman. I conclude that there never was invented an artifice so excellent and marvellous and of such genius." This was the oldest documented astrarium, but it was not the first. Archimedes was said to have made one, and the most famous example is the Antikythera mechanism.

ABOVE: Richard of Wallingford pointing to a clock, in reference to his gift to St Albans Abbey in England, a clock designed to regulate the working day of the abbey's monks.

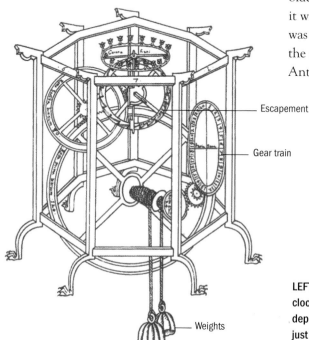

— Escapement

— Gear train

— Weights

The first
me(c)hani(c)al
cl(o)cks
were built 700 years ago

LEFT: A tracing of an illustration from Giovanni de Dondi's 1364 clock treatise, *Il Tractatus Astarii*, shows his astrarium. It doesn't depict the upper section with its complex system of wheels, but just the main gear train, weights, and escapement mechanism.

The Antikythera mechanism

In October 1900, a boatload of sponge-divers took refuge from a storm near the island of Antikythera, 50km (30 miles) northwest of Crete. When the storm died down, they noticed a shipwreck on the seabed 60m (200ft) down. They dived and recovered a number of statues, whose elegance suggested that the ship had been carrying stolen artefacts from Greece towards Rome. Apart from the bronze statues, they brought up what at first sight looked like a lump of rock with an embedded gear wheel.

The object is immensely corroded, but has been examined many times by experts, using X-rays and other probes, and is now known to be an astronomical analogue calculator or orrery. Apparently made around 100 BC, it contained somewhere between 30 and 72 cogwheels, all cut with triangular teeth from a single sheet of bronze.

When a date was entered by means of a crank, the machine would have calculated the positions of the Sun, the Moon, and the planets. There seems to have been a set of instructions in Greek engraved in bronze. There were three dials to show the results, and the calendar dial could be moved to allow for an extra day every fourth year, even though the idea of leap years was not formally invented until many decades later. The Antikythera mechanism is still being studied. It was not, strictly speaking, a clock, and had no source of power, but it is the oldest known mechanical calculator, and an amazing tribute to the mathematical and mechanical skill of the ancient Greeks.

ABOVE: This corroded cogwheel embedded in rock is the surviving fragment of a sophisticated ancient calculating device, the Antikythera mechanism.

A bronze astronomical calculator,
THE ANTIKYTHERA MECHANISM
was built in about 100 BC

LEFT: This reconstruction shows how the Antikythera mechanism may have looked. Its dozens of cogwheels calculated the positions of the Sun, Moon, and planets at any given time.

OPPOSITE: The astronomical clock on the Old Town Hall in Prague dates from 1410 and is the oldest one still in working order. The dial is in the form of an astrolabe, showing the Sun and Moon in the sky.

The Roman military engineer and writer Vitruvius, who served in the army under Julius Caesar, wrote ten books of architecture around 2000 years ago. In the last of them he describes many machines, including the ballista and the Archimedes screw, but he describes only the most basic of gears, and does not mention a machine with anything like the complexity of the Antikythera mechanism, even though that had been made a hundred years earlier, which suggests that it must have been extremely rare, if not unique.

Turret clocks

The medieval clocks of the great cathedrals of Europe were the forerunners of dozens of clocks built from 1500 onwards in many great towns and cities, usually in the towers of churches or town halls, and commonly known as tower clocks or turret clocks.

King Henry VIII brought a group of French Huguenots over to England to sort out the clocks in his various palaces, and this was the beginning of a gradual migration of Huguenot clockmakers from mainland Europe to England, where they could escape religious persecution. Clockmaking became widespread during the late 16th and early 17th centuries, in France, Flanders (Belgium), and the Netherlands, but especially in England.

CLOCK FUNDAMENTALS

Every clock has three basic constituents: a power source, a counting system with a display – bells, hands, or digital output – and a regulator to connect the other two components. Modern clocks, for example, have batteries for power, a digital display, and a quartz oscillator to connect the two. In water clocks, the power source is the falling water, the display may be a rising float with a pointer and a scale of hours, and the regulator is the hole through which the water has to fall. Turret clocks were powered by falling weights – usually big lumps of iron. The weights hung on the ends of a rope, which was wound round a cylinder, so that as one weight dropped the rope turned the cylinder, which through cogwheels turned the rest of the gear train, and finally the hands. At the hour, the pull of the second weight drove the hammer to strike the bell.

Without any restraint, the weights would fall faster and faster, and the hands would spin round the dial at high speed. The check was the escapement mechanism (as shown on the right), which held back one of the cogwheels, and allowed one tooth to escape past the barrier with each oscillation of the regulator.

The early turret clocks had regulators (see "Clock Fundamentals" on the opposite page) known as verge and foliot. This was an escapement mechanism, which allowed the clock to control its rate by advancing the gear train at regular intervals.

The verge was a bar whose pallets interrupted the turning of the escape wheel, or crown wheel. It was connected to a foliot, which was a horizontal bar with a large mass at each end. These masses gave the foliot considerable inertia, so that to swing back and forth took a second or two. The speed of this slow oscillation, which controlled the movement of the escape wheel, could be adjusted by sliding the masses back and forth along the bar.

This development of a mechanical oscillator (earlier clocks used water or mercury as a regulator) was a big step forward, and allowed clocks to run with much greater precision. However, verge-and-foliot clocks could vary according to the weather, and they could easily be an hour or more wrong by the end of each day. What was needed was a more reliable device to count the seconds. The first person to come up with such a device – the pendulum – was Galileo Galilei.

RIGHT: The clock inside St Mark's Clocktower in Venice was originally built with a verge-and-foliot regulator. This was later replaced with a pendulum.

Galileo di Vincenzo Bonaiuti de' Galilei was born in Pisa on 15 February 1564. He became a world-famous scientist: first, he successfully disputed the views of Aristotle on the science of falling objects; then he used a telescope to study the heavens, and discovered the moons of Jupiter. His support for the Copernican theory that the Sun, rather than the Earth, was the centre of the Universe got him into trouble with the Church, and he spent the last 15 years of his life under house arrest.

Galileo and the pendulum

GALILEO WAS ONE OF THE FIRST scientists in the West not only to challenge the wisdom of the ancients but also to do experiments to test his theories. While he was a medical student in Pisa in 1582, so the story goes, Galileo went as usual to the *duomo* – the cathedral beside the leaning tower. The architecture is stunning, and the paintings are breathtaking, but he had seen them a hundred times, and meanwhile the sermon was long and tedious. He noticed idly that the great bronze lamp was swinging on its long chain, driven by the draught from the door. To occupy his mind, he timed it, using his pulse as a measure.

ABOVE: In the last years of his life, Galileo Galilei (1564–1642) realized that pendulums could be used to keep time.

At first the lamp was barely moving, just a few centimetres each way, and he noted that it took nine or ten pulse beats to go to and fro. Then someone went out and left the door open, and the lamp, caught by the cold blast, swung in wider and wider arcs, until it was moving at least 2m (6ft) back and forth. He timed it again, and was surprised to note that it still took nine or ten pulse beats for a complete swing. (There is still a great bronze lamp hanging from a chain in the *duomo*, but alas it is fastened up against the wall and can no longer swing in the draught.)

Back at home, Galileo hung a weight on a piece of string, and began investigating the behaviour of this simple pendulum. First, he confirmed his observation in the *duomo*, that the pendulum always takes the same time for each swing, regardless of how far it swings. Second, he showed that the mass of the weight does not matter – a 200g (7oz) weight swings at the same rate as a 100g (3.5oz) weight. However, he found that the length of the string does matter: with a longer string, the pendulum swings more slowly. The time or "period" of the swing is proportional to the square root of the length, so a pendulum four times as long takes twice as long for each swing.

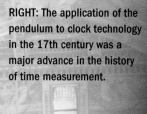

RIGHT: The application of the pendulum to clock technology in the 17th century was a major advance in the history of time measurement.

Galileo used this knowledge to make a small portable pulse meter, a weight suspended on a string from a wooden stand some 20cm (8in) high, arranged so that he could vary the length of the string. As a medical student, he used it to time the pulses of his patients, adjusting the length of the string until, when he gave it a push, the pendulum swung just in time with the pulse. Then he could read off the pulse rate from the wooden stand, which probably just said "fast", "medium", or "slow", since he had no precise clock to measure with. The medical authorities saw how useful this was, and stole the idea, without giving him any credit.

Galileo realized that the regularity of the pendulum's swing would make it a good regulator for a mechanical clock. He designed such a clock, in 1637, but this was not until near the end of his life, when he was confined under house arrest to his villa at Arcetri, and gradually going blind. As a result, he never built it. His son, Vincenzio, partly constructed the clock in 1649, but also did not live to complete the project. The first pendulum clock was constructed a few years later by the Dutch scientist Christiaan Huygens, building on Galileo's work.

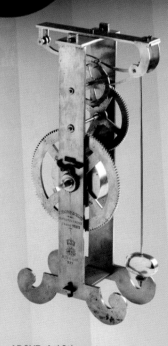

ABOVE: A 19th-century reconstruction of a pendulum clock Galileo designed in 1637.

Christiaan Huygens (1629–95) was a polymath with an impressive range of skills. He was the first astronomer to decipher the shape of Saturn's rings. He also discovered Titan, Saturn's largest moon. He was the first physicist to propose a wave theory of light and to show how the waves could explain such phenomena as refraction. He also published a book on probability theory.

BELOW: In this 1955 painting by Hugh Chevins, Huygens is seen in the clockmaker's shop of Salomon Coster in 1657, holding his newly built pendulum clock.

THE SCIENTIFIC REVOLUTION

WESTERN SCIENCE WAS FLOURISHING BY THE MIDDLE of the 17th century. Huygens was elected a fellow of the Royal Society in London in 1663, only three years after it was founded. In 1666 he took up a post at the brand-new Académie des Sciences in Paris. In other words, he was an exceptional scientist; he was also well-respected in European scientific society.

In 1657, Huygens designed the world's first pendulum clock, and had it built by Salomon Coster, a clockmaker in The Hague. Following Galileo, Huygens realized that a pendulum is a regular timekeeper, and therefore would be more reliable than a foliot in an escapement mechanism. A much better mathematician than Galileo, Huygens worked out the formula for the time of the complete to-and-fro swing.

HUYGENS' FORMULA

Huygens' formula for the time period (complete to-and-fro swing) of a pendulum looks like this:
$$T = 2\pi\sqrt{(l/g)}$$
l is the length of the pendulum and g is the gravitational constant, 9.81m/s/s (32.18ft/s/s).
For a pendulum 1m (3.28ft) long, the time is just over 2 seconds.

A LEAP FORWARD FOR SCIENCE

Pendulum clocks remained the most precise in the world for nearly 300 years, and changed the face of science in general and astronomy in particular, because all observations and measurements could now be underpinned by precise measurement of time.

Hampered by the verge escapement, Huygens had to make his pendulum swing through a wide arc, and as a result had to invent new mathematics to describe what he needed (published in 1673), and then make special "cheeks" to control the shape of the swing. Nevertheless, he made the most precise clocks ever seen until that date, and claimed a precision of around 10 seconds per day (which was probably a hundred times better than the verge-and-foliot clocks), although in practice it was probably more like a minute a day.

Huygens almost certainly knew that a long pendulum would keep better time, but because he had to use an arc of around 90 degrees (45 degrees each side of the vertical) he had to keep his pendulum short if the clock was not going to occupy the whole room.

A Norwich-born clockmaker who had moved to London in 1631, Ahasuerus Fromanteel set up his business in Southwark, south London. He sent his son John off to be apprenticed to Salomon Coster in The Hague in 1657, and when he returned in May 1658 John brought back the secret of the pendulum. His father jumped on it, and in October of that year placed an advertisement in the London newspaper *Mercurius*: "There is lately a way found out for making Clocks that go exact and keep equaller time than any now made without this Regulator… Made by Ahasuerus Fromanteel, who made the first that were in England." Mind you, although he made longcase clocks, he was still fitting them with the short pendulum pioneered by Huygens.

Greater precision

The next step was the invention of the anchor escapement, by Joseph Knibb, William Clement, or Robert Hooke – there is some controversy about who actually did it. Knibb was an Oxford clockmaker with a workshop in Holywell Street, from where, early in 1670, he made for Wadham College (just round the corner) a clock

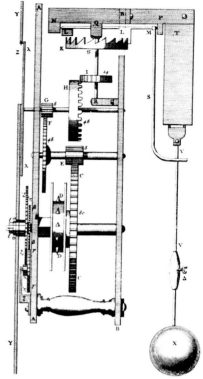

ABOVE: Huygens' design for the first pendulum clock, as described in his book *Horologium* (1658).

The time **swing** of a pendulum depends only on its **length**

RIGHT: The elegant face of a walnut longcase clock, with calendar date, designed by Joseph Knibb in 1675.

THE ANCHOR ESCAPEMENT

The anchor escapement was so called because the C-shape of the arms carrying the pallets looks rather like a miniature anchor. The mechanism was important, not because it was inherently more precise than the verge, but because it allowed a much smaller swing of the pendulum (12° instead of 90°), and therefore made it possible to use a long pendulum, and put it in a "longcase" clock.

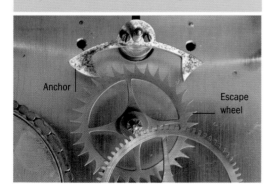

Anchor

Escape wheel

with an anchor escapement; later that year he moved to London and became clockmaker to King Charles II.

Clement was a well-known London-based clockmaker who started using anchor escapements in 1671. Hooke was curator of experiments at the Royal Society, and a lifelong coffee-parlour frequenter with architect Christopher Wren and Edmond Halley (the Astronomer Royal who recognized the returning comet that is now named after him – Halley's comet). Hooke was a brilliant scientist, and interested in clocks, but he was not a clockmaker, so it seems likely that Knibb was responsible for the new escapement.

The greatest precision in the 17th century was achieved by Thomas Tompion, who was baptized in Bedfordshire, set up business as a clockmaker in London, and became the most famous of his age. In 1675, when the Greenwich Royal Observatory was built, he was commissioned by Sir Jonas Moore (who had supervised the building) to make two "regulators" (precision clocks) for the use of the first Astronomer Royal, John Flamsteed.

RICHARD TOWNELEY

Richard Towneley (1629–1707) of Towneley Hall in Lancashire was a mathematician, astronomer, and wide-ranging scientist. Not only did he design the dead-beat escapement for Tompion's Greenwich "regulators", and build a third "regulator" himself, but he was also the first person in Britain to record rainfall regularly, and reported that much more rain fell on Lancashire than on Paris.

On 27 April 1661, he and his physician Henry Power climbed 300m (980ft) up nearby Pendle Hill carrying a mercury barometer, recently invented by Evangelista Torricelli. They discovered that the air pressure (P) was lower at the top of the hill, and that the volume (V) of a fixed amount of air increased. He wrote to his friend Robert Boyle about this, and the relationship (PV = constant) eventually became known as Boyle's law. Robert Boyle, however, called it "Mr Towneley's hypothesis".

Tompion and Flamsteed agreed that these "regulators" should incorporate the newly invented "dead-beat" escapements, which unlike the anchor escapement did not cause the escape wheel to recoil or bounce back with each beat. These escapements were designed by Richard Towneley, and later improved by Tompion.

The pendulums were 4m (13ft) long, so they took 2 seconds to swing each way. With the dead-beat escapement, however, they had to swing from front to back, rather than from side to side, so hollows had to be made in the wooden panelling to accommodate the back swing. These "regulators" ran for a whole year before they needed rewinding, and even then continued to run while they were being wound. They kept time to within 2 seconds a day, which made them the most precise clocks in the world at the end of the 17th century.

BELOW: Thomas Tompion is regarded as the Father of English clockmaking. Today, his surviving clocks command huge prices at auction.

Thomas Tompion's clocks of 1676 were precise to within 2 seconds a day

The first ever portable, mechanical timepiece worn or carried was probably made in 1510. Henry VIII came to the throne in 1509, and may have had a watch – the circular medallion that hangs round his neck in various portraits may be the back of a watch. Otherwise, in the early 1500s watches existed only in Germany: most people in England relied on church clocks.

The first watches

RIGHT: An 18th-century advertisement for the "Improved Pedometer, or Waywiser" shows a fashionable London gentleman sporting a pocketwatch hanging from the belt of his breeches.

THE FIRST WATCHES IN ENGLAND were made by the newly arrived Huguenot artisans. By 1600, a defined English style of watch emerged and London soon became the clock and watch centre of the world. In those days, watches were worn on a chain round the neck, and had only an hour hand, since they were so imprecise they might lose or gain an hour a day.

The word "watch" originally meant a portable timepiece that did not strike the hours, and was even applied to large machines that only showed the time, but did not strike; a "clock" was a timepiece that struck a bell. Techniques were steadily improved over the years, and in the early

1700s, London clockmaker George Graham produced a pocket watch that was precise to within a minute a day.

No watch could accommodate a pendulum, so the regulators were balance wheels, which oscillated first one way and then the other, always pulled back towards the central position by a helical balance spring. Balance wheels were actually improved versions of the foliot – there is one in the drawing of Giovanni de Dondi's 1364 clock (see p177) – and were given a great boost by John Harrison's experience with the watch made for him by Jefferys (see p196), and by his own H4 (see p197). Balance wheels were much improved as new and better steel became available, and they are still used in some mechanical watches today. However, they were largely replaced by the quartz oscillator (see p200) in the 1970s.

Self-winding watches were invented by Swiss watchmaker Abraham-Louis Perrelet in 1770. They have a small eccentric weight that turns on a pivot. Each time the wearer moves their arm, the weight swings, and whichever way it swings it winds the mainspring a little. There is a clutch to prevent over-winding. The idea has not changed greatly since the 18th century: self-winding watches still have the eccentric weight, which now drives a tiny generator which in turn charges up the battery.

ABOVE: A trench watch from World War I, without its wrist strap. Like a pocket watch, trench watches had hinged covers. Until 1914, wristwatches were regarded as effeminate. Only in the exigencies of the trenches did officers appreciate the convenience of a wristwatch.

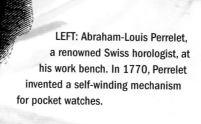

LEFT: Abraham-Louis Perrelet, a renowned Swiss horologist, at his work bench. In 1770, Perrelet invented a self-winding mechanism for pocket watches.

By the 17th century, exploring, trading, and raiding by sea had become important for many maritime peoples, and to be efficient and safe, sailors had to be able to find out where they were, even when there was no land in sight.

THE PROBLEM OF LONGITUDE

ABOVE: A sailor uses a quadrant to calculate his latitude from a prominent star above the horizon. Such calculations are relatively easy.

ANY POSITION ON EARTH CAN BE PINPOINTED by two coordinates, latitude and longitude. Your latitude tells you how far you are north or south of the equator. So the equator itself is latitude 0 degrees. Rome (Italy), Madrid (Spain), New York (USA), Beijing (China), and Pyongyang (North Korea) are all close to 40 degrees north, while Melbourne (Australia), Valdivia (Chile), Bahia Blanca (Argentina), and Wellington (New Zealand) are all close to 40 degrees south. Finding out your latitude is fairly simple. All you need to do is measure how far the Sun is above the horizon at noon, when it is at its highest point in the day. At the spring and autumn equinoxes, if you measure the angle of the Sun above the horizon as x°, then your latitude is x°. For the rest of the year you need tables to tell you how high the Sun should be. At night you can use the angle of a prominent star – in the northern hemisphere Polaris, the pole star, is a good one.

To measure LONGITUDE you need a very accurate clock

SIDEREAL TIME

For centuries, most people have used solar time, the day being the average time between successive sunrises. Astronomers prefer to use "sidereal" time based on the stars: the word sidereal comes from the Latin *sidus*, meaning "star". The interval between the successive rising times of any particular star is one sidereal day, which is equal to about 23hrs 56mins 4.1s. This differs from the solar day of 24 hours because the sidereal day depends only on the rotation of the Earth, relative to the distant stars. The solar day is longer, because it depends both on the rotation of the Earth and on the movement of the Earth in its orbit around the Sun. In other words, reckoning from yesterday's sunrise, the sidereal day is over just before the Sun comes back into sight.

In the southern hemisphere, it is not so simple, but there is a faint star called Sigma Octantis roughly 4.5 times the length of the Southern Cross in a line from the long end of the Southern Cross.

Finding your longitude is much more difficult. Sailors always had trouble finding out how far east or west they were. There are neither landmarks nor simple heavenmarks. Neither Columbus nor Cabot had much of a clue about how far away the next land was when they sailed optimistically west from Europe at the end of the 15th century, and for almost 300 years after that sailors had to use dead reckoning – estimating how fast they were going relative to the sea – to have an idea where they were.

The crunch came, literally, on 22 October 1707 when a fleet of British navy ships was returning from Portugal, and as a direct result of a navigational error they sailed at full speed into the Scilly Isles. Four ships sank. Nearly 2000 men died. The Admiral of the Fleet, Sir Cloudesley Shovell, was thrown onto the beach, badly injured, and was then murdered by a woman who stole the emerald ring from his finger.

ABOVE: A 1765 globe divides the world into latitude and longitude. Without an accurate means of determining longitude, nautical navigation was a highly inexact science.

A reward offered

This was one of the greatest naval disasters in British history. The government and Queen Anne decided that serious efforts had to be made to solve the problem of longitude. So in 1714 an act of parliament was passed offering a prize of £20,000 (worth several million today) to anyone who could do so, or specifically to anyone who could find a way of determining longitude to within 30 miles (48km). The prize was to be administered by a specially appointed Board of Longitude.

Many people came forward with ideas, some of which were more sensible than others, and the Board awarded significant amounts of money to more than a dozen of them. Some were astronomical suggestions, which tended to find favour with the Board, several of whom were astronomers. The Greenwich Observatory itself had been set up specifically to help navigation. Other people, however, thought that the solution would come from a really precise clock. Wherever you are in the world you can measure local time, for the Sun is always highest in the sky at 12 noon. If you also know simultaneously what time it is at Greenwich, then you can immediately work our how far you are west or east of Greenwich. Suppose, for example, your local time is 12 noon but your precise clock, set to Greenwich time, says 5pm, then you know you are 5 hours behind Greenwich, which means 5 hours west. Since each hour corresponds to 15 degrees (see p156) you must be at longitude 5 x 15 = 75 degrees west, which puts you near the eastern coast of America.

There already existed many clocks that were easily precise enough – Tompion's Greenwich clocks, to name but two – but they were all controlled by pendulums, and pendulums don't work properly on board a ship, rolling about in the waves. Many clockmakers tried to make clocks that would be truly portable, and would also remain precise in hot and cold climates. By far the most successful of them was a village carpenter called John Harrison.

John Harrison was born at Foulby, near Wakefield in Yorkshire in 1693. His father was a carpenter, and John took up the same career, making doors, windows, and furniture for anyone who could afford to pay, and in particular for Brocklesby Park, the estate of the Earl of Yarborough.

JOHN HARRISON

HARRISON BECAME FASCINATED BY CLOCKS at an early age, and in his spare time repaired clocks that were broken or misbehaving. From the age of 20, he began to build longcase clocks, and three of his early clocks survive. They are made almost entirely of wood: in particular all the gear wheels are wooden, with teeth cut from oak, cunningly arranged so that the grain of the wood is radial, running out along each tooth.

In the early 1720s, Harrison was commissioned to make a clock for the stable block at Brocklesby Park. He made this clock also entirely of wood, and finished it in 1722, but it frequently went wrong and he kept having to go back to fix it. While doing so he incorporated several new features, including a "grasshopper" escapement, an entirely new idea to eliminate friction.

He found that the oil usually used for lubrication tends to thicken in winter, thin in summer, and generally cause problems. He therefore made all his bearings of a special hardwood called lignum vitae, which feels greasy and is effectively self-lubricating. After almost 300 years, the Brocklesby Park clock is still ticking, and keeps remarkably good time, losing perhaps a minute a week. It has not been lubricated in all that time.

Harrison visited London in about 1728, looking for support for his work. He went first to Greenwich, where the Astronomer Royal, Edmond

BELOW: John Harrison poses for his portrait holding the ingenious pocket watch made for him by the watchmaker John Jefferys. It was this watch that gave Harrison the idea for his sea watch H4.

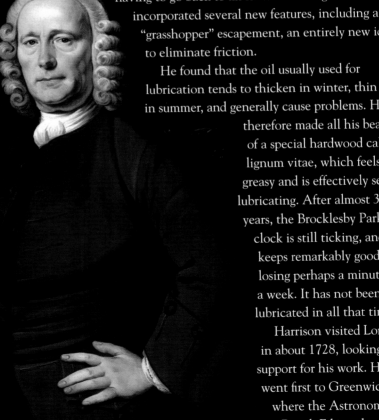

...fancy, sent him on to meet George Graham, the foremost clockmaker in London, who had worked with Tompion. To begin with it was an uncomfortable meeting, between the rough Lincolnshire carpenter and the refined Fellow of the Royal Society. Harrison wrote later: "Mr Graham began, as I thought it, very roughly with me… which … occasioned me to become rough too; but however, we got the ice broke… and indeed, he became as at last, vastly surprised at the Thoughts or Methods I had taken." The meeting began at 10am, and continued through the day and over dinner that evening. Graham even offered to lend Harrison money for research.

Harrison decided to have a serious attempt at winning the longitude prize, and in fact he spent most of the rest of his life on this single project. First he built two longcase clocks, known as "Precision Pendulum 1" and "Precision Pendulum 2" or PP1 and PP2 for short. These already had all the refinements that he had built into his Brocklesby Park clock, but he needed them for further research.

He knew that temperature control was important. All clocks with simple pendulums run faster in winter than in summer. This is because the metal pendulum rod expands slightly in hot weather, and as it gets longer each swing takes slightly more time, so the clock will run more slowly. The problem would clearly be much worse on board a ship that might sail from the Arctic Ocean to the tropics.

His solution was the gridiron pendulum, which in its simplest form comprises three iron rods hanging down and two brass rods sticking up. When the temperature increases the iron rods expand, but the brass rods expand slightly more, so that the overall effective length of the pendulum is unchanged. In fact, Harrison found it

ABOVE Harrison's eight[...] clock made in 1715, w[...] he was 22. The mecha[...] is made almost entirel[...] of wood.

In 1714 the British government offered
a prize of **£20,000**
(£10 million in today's money) to anyone
who could solve the problem of longitude

THE GRIDIRON PENDULUM

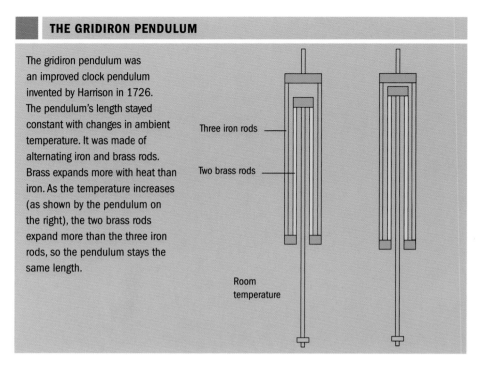

The gridiron pendulum was an improved clock pendulum invented by Harrison in 1726. The pendulum's length stayed constant with changes in ambient temperature. It was made of alternating iron and brass rods. Brass expands more with heat than iron. As the temperature increases (as shown by the pendulum on the right), the two brass rods expand more than the three iron rods, so the pendulum stays the same length.

Three iron rods

Two brass rods

Room temperature

better to use five iron rods and four brass ones. (At about the same time, George Graham invented a pendulum containing a tall, thin flask of mercury, which had the same effect: as the iron expanded downwards, the mercury expanded upwards, leaving the effective length unchanged.)

Tried and tested

Few clockmakers conducted the same sort of rigorous scientific research as Harrison. He put one of his PP clocks in a cold room with an open window and the other in a hot room with a roaring fire, compared them for a few days, and then swapped them over, making adjustments when necessary. Checking his clocks for precision was not easy: there was no available standard like the Greenwich time signal. What he did was to watch every night from his window and note exactly when a particular star disappeared behind his neighbour's chimney. This gave him a precise measure of 24 hours, to within a twentieth of a second, which he used to calibrate his clocks.

By using such cunning scientific methods to improve the timekeeping, Harrison succeeded in getting his PP clocks – he called them "timekeepers" – running precisely to within one second a month, which was far better than any clock in London. What is more, with their wooden cogs and lignum vitae bearings, they needed no lubrication.

Harrison then set out to make sea clocks, knowing that he could not use a pendulum, and relying instead on pairs of oscillating springs. The first of these, completed in 1735 and later called just "H1", was perhaps the most spectacular. It was 60cm (2ft) high and 90cm (3ft) wide, with a frame of brass, but gear wheels of oak, as in Harrison's previous clocks. It was powered by a spring, and had to be wound every day. The four dials showed hours, minutes, seconds, and the date.

Chasing the prize

In 1736, H1 was sent by the Board of Longitude on a trial voyage to Portugal. The weather was rough and Harrison was dreadfully seasick. The clock performed moderately well, and even better on the return voyage. In fact the ship might have been wrecked without H1, for Harrison warned the captain they were 100 km (60 miles) farther west than his officers' navigation had suggested. The Board decided that H1 had not done well enough to receive the prize, but for the first time they realized that a good clock might solve the longitude problem, and they awarded Harrison £250 for further research, and promised another £250 if he could make a better machine. That same year, Harrison moved to London.

He went on to produce two more huge sea clocks, H2 (1741) and H3 (1759), which did not perform quite well enough to win the prize, though the Board continued to support him financially. Then came what may have been a stroke of luck, or perhaps another stroke of genius: he designed a watch. Until then clocks had been far better timekeepers than watches, which was why Harrison originally set out to build sea clocks.

He was much taken with a pocket watch made by George Graham's successor, Thomas Mudge, and he designed one himself. Not being a watchmaker, he asked watchmaker John Jefferys to make it for him. Clock- and watchmaking had diverged, because making watches was much more intricate, and demanded separate techniques: the cogwheels could not be wooden, for example. Because

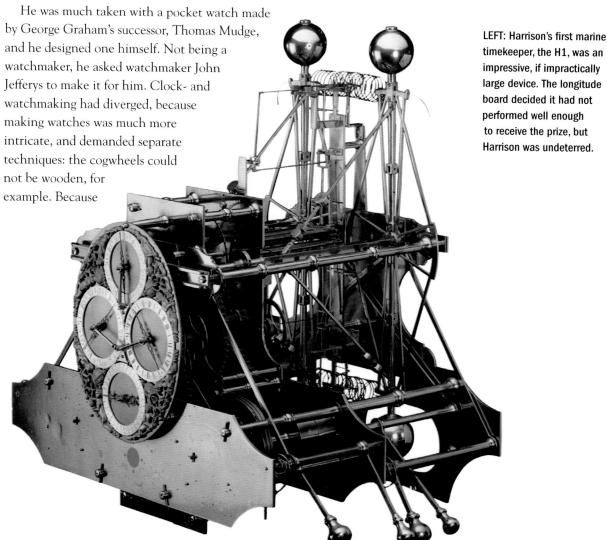

LEFT: Harrison's first marine timekeeper, the H1, was an impressive, if impractically large device. The longitude board decided it had not performed well enough to receive the prize, but Harrison was undeterred.

ABOVE: A copy of Harrison's H4 sea watch was taken on Captain James Cook's voyage round the world in 1768.

of the difficulties of miniature engineering without reliable steel for the springs, most watches were far less precise than clocks.

Harrison was agreeably surprised by the performance of his Jefferys watch (which he is holding in his hand in the well-known portrait on p192), and he realized that a watch might be the final answer to the longitude problem. So he set out to make a "sea watch", and in 1759 produced H4, a large pocket watch 13cm (5.2in) in diameter.

This was an extraordinary instrument. H4 was more precise than any other timekeeper, and after its first trial voyage to Jamaica was found to be in error by only 5 seconds, which corresponds to an error of just 1 nautical mile (1.85km; 1.2 miles). Later a copy of H4 went with Captain James Cook on his first voyage round the world. When Cook got back to Plymouth after three years at sea, the watch showed an error of only 13km (8 miles).

Still the Board of Longitude were reluctant to award Harrison the prize. In the end he appealed directly to King George III, who told them to stop being so mean.

Harrison's H4 **sea watch** went on a trial from England to Jamaica, and was only 5 seconds wrong at the end of the voyage

THE FIRST AMERICAN CLOCKS

The American clock industry really began in 1797, when an inventor in Connecticut, Eli Terry Sr (1772–1852), took out his first clock patent. He introduced the idea of mass production to the process of making clocks: this brought down the price and made them available to average American citizens. In order to do this he bought a mill, and used water power to allow his apprentices to rough-cut wooden gear wheels, which were then hand-finished by skilled journeymen clockmakers. He also devised jigs and methods that allowed him to make interchangeable parts for his clocks, which greatly speeded up production.

A skilled clockmaker could perhaps make 10 clocks in a year, but in 1806 Terry signed a contract to produce 4000 wooden clock movements – other people would make the cases. He spent two years perfecting the machinery and the production line for his new shelf clock – one of the first machines in the world to be mass-produced using interchangeable parts – then in the third year he produced 3000 clocks.

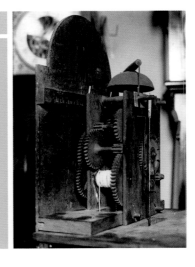

Eventually they gave Harrison a large payment, which brought the total up to just over £23,000, some of which had been paid to him decades earlier to allow him to keep working on the problem. So although he did not technically win the prize, Harrison did well financially, and he was proud of what he had achieved with H4: "I think I may make bold to say, that there is neither any other Mechanical or Mathematical thing in the World that is more beautiful or curious in texture, than this my watch, or Timekeeper for the Longitude… and I heartily thank Almighty God that I have lived so long, as in some measure to complete it." He died on his 83rd birthday, in 1776.

Several clock- and watchmakers picked up the gauntlet after Harrison died and the details of H4 had been published. The most successful was John Arnold, who created the first affordable, precise, portable timekeeper, and in 1779 introduced the word "chronometer" to describe it. He was followed by Thomas Earnshaw, who further simplified and standardized the chronometer. By 1800 he had produced a superb instrument that could be mass-produced, and it barely changed for 150 years. Both Arnold and Earnshaw received money from the Board of Longitude for their work. All chronometers for the British navy were tested in a particular room at Greenwich, at room temperature, in an oven, and in a fridge.

BELOW: Harrison's brilliantly elegant H4, shown at actual size.

Later pendulum clocks

In 1889, German instrument-maker Sigmund Riefler patented a new escapement mechanism, in which the impulse to keep the pendulum swinging was imparted not to the pendulum but to the flexible metal strip that supports it. Riefler clocks became the most precise (to 10ms per day) in the world. The first US time standard was provided by the Bureau of Standards (now NIST) between 1904 and 1929 using Riefler clocks, and time signals were delivered by telegraph wire.

In 1921, British railway engineer William Hamilton Shortt built a clock with two pendulums, which was not a new idea. The master pendulum swung in a sealed copper vacuum tank and kept highly precise time, and every 30 seconds made very brief contact with the escapement. This transmitted an electromagnetic impulse to the slave pendulum, which stood a few metres away and actually controlled the movement and the electrical contacts needed to provide power. This was the first clock that was sufficiently precise (to about 1 second per year) to detect small seasonal changes in the rotation of the Earth.

THE CONGREVE CLOCK

William Congreve was a British inventor and soldier who developed artillery rockets for use in warfare, after they had been used against the British in India. In 1808, Congreve designed a clock in which the regulator was not a pendulum but a ball rolling down a zigzag slope. When it reached the end, the ball tripped the escapement, the hands moved, and the plane tilted so that the ball ran back the other way. The journey of the ball took 30 seconds, and Congreve hoped that, with such a long period, as opposed to the 2-second pendulum, his clock could be very precise. Unfortunately it turned out to be rather unreliable: there were problems with friction and probably with dust in the grooves. However, Congreve presented his first clock to the Prince of Wales. In fact, Congreve was not the first: rolling-ball clocks had been invented at least 100 years before, but his version became better known.

BELOW: One of Alexander Bain's electrical clocks, dating from about 1850.

Writing in *The History of Clocks and Watches* (1999), Eric Bruton said, "Most new ideas are unoriginal, unworkable, uneconomic, or unscientific. On the very, very rare occasion, an idea emerges that is really unique and brilliant, such was the use of electricity." He was introducing the idea of electric clocks, which began in the early 1800s and came into sharp focus with Alexander Bain in 1837.

ELECTRIC CLOCKS AND WATCHES

BAIN HAD THE IDEA OF USING ELECTRICITY not merely to push a clock's pendulum but also to provide power to drive the clock, rather than using weights or a spring. He also managed to convey signals through a telegraph wire to another clock some distance away, so that their two pendulums would keep exactly in sync; they would be master and slave. He managed to place a master clock at Edinburgh railway station and a slave at Glasgow railway station some 80km (50 miles) away, so that the two remained perfectly in time with one another.

This system was used to link Greenwich to the clock in the Palace of Westminster, whose bell is called Big Ben, and to the Post Office, which then became a major distributor of time. Electric clocks became increasingly common during the 19th and 20th centuries. Once a stable mains electricity system was in place, clocks could be run according to the frequency of the alternating current. Most countries use either 50 or 60Hz: ie the current flows first one way and then the other 50 or 60 times per second. Knowing this, manufacturers can make clocks that simply count the oscillations and translate that into seconds. This relies on power generators delivering a

ALEXANDER BAIN

"Sandy" Bain was born in 1811 at the village of Watten, in the far north of Scotland, many miles from any town. He was hopeless at school in the winter, and hopeless as a shepherd in the summer, because he was always dreaming, usually about clocks. He became apprenticed to a clockmaker, and then became intrigued by the newly available electricity. His first major invention was an electric clock, which relied on a pendulum as a regulator, but used a solenoid and a magnet to give the pendulum a push every swing. Desperate for money to develop it, he showed his plans to the distinguished Professor Charles Wheatstone, who had already patented the electric telegraph with William Fothergill Cooke, and asked for his opinion. Wheatstone said, "Oh, I shouldn't bother to develop these things any further! There's no future in them." A few weeks later, Wheatstone demonstrated his own electric clock to the Royal Society. Bain's most amazing invention was the fax machine, which he patented in 1843, 30 years before the telephone was invented, and about 130 years before faxing really came of age. Sadly Bain was financially incompetent, and died poor in 1877.

The advantage of mains **electricity** was that all clocks on the same circuit would show the same time

highly regular output. One advantage of this system is that all the clocks on one circuit will stay exactly in sync; so all the clocks in one company, for example, once they have been set up properly, will always show exactly the same time.

BELOW: This diagram, drawn by Bain in 1840, shows a series of clocks on the same circuit, showing the same time.

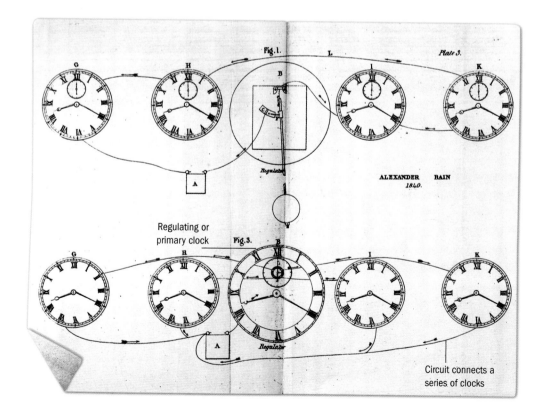

Regulating or primary clock

Circuit connects a series of clocks

ABOVE: The quartz crystal is cut by laser to the shape of a tiny tuning fork.

Quartz oscillators

Quartz is a mineral, chemically the same as sand (silicon dioxide, SiO_2). It is the second most abundant mineral in the Earth's crust, after feldspar, and is found in various crystalline forms, several of which are attractive, and some of which are semi-precious stones, including agate, amethyst, and onyx.

Apply an alternating voltage to a quartz crystal (from a battery and some circuitry) and it will rock to and fro. When the crystal is cut to the correct size and shape, the alternating voltage will set up natural vibrations in the crystal, and these can be tuned to any desired frequency.

Manufacture is easier for small crystals of quartz, which vibrate at very high frequencies. The crystal is usually cut to the shape of a tiny tuning fork, and trimmed by a laser until it vibrates with a frequency of 32,768Hz (cycles per second). This is 2^{15}, or 2x2x2x2x2x2x2x2x2x2x2x2x2x2x2, and it is a good number because a 15-bit binary digital counter (which counts in twos) can easily count this number of oscillations, and then advance the time on a watch by just one second. Such quartz crystals oscillate with an extraordinarily precise frequency, and are not significantly affected by variations in temperature.

The quartz resonator has therefore largely taken over from the pendulum and the balance wheel, and because it comes in a flat package about 4mm (1/6in) long, it can easily be fitted inside even a small watch.

Extraordinary precision

Because of the precise frequency of the vibrations, quartz clocks and watches have a precision of around six parts per million, which means they will stay within half a second per day under normal conditions. If the clock is compared to an atomic clock and tweaked accordingly, the precision can be improved to be within 10 seconds a year.

Quartz crystal oscillators were developed in the early 1920s by Walter G Cady and Warren Marrison in the US and David William Dye at the National Physical Laboratory in England. The National Bureau of Standards in the US based the US time standard on quartz clocks between the 1930s and the 1960s, and in the late 1960s, when semiconductor digital logic (which allows the

PIEZOELECTRICITY

Quartz crystals exhibit piezoelectricity, which means that if squeezed they develop an electrical potential – a voltage between one side and the other. This was discovered in France in 1880 by Jaques Curie and his younger brother Pierre, who married radioactivity pioneer Marie Sklodowska in 1895, and shared with her and Henri Becquerel the 1903 Nobel Prize for physics. At the end of the 19th century, quartz crystals (below) were used in phonograph pick-ups. The varying pressure from the needle as it rode up and down over the bumps and hollows in the record produced a varying voltage that was used to make sound. Jacques and Pierre also discovered the reverse piezoelectric effect: that applying a voltage to a quartz crystal causes it to change shape.

crystal to communicate with the digital counter) had become cheap, the
first quartz watches were produced – the pioneer being Seiko's Astron. This
watch was launched on Christmas Day 1969, after 10 years of development.
Within the first week, 100 watches had been sold, each costing 450,000 yen
(£3300). This was the same price as an average car.

During the last 30 years, quartz crystal oscillators have invaded not only
ordinary clocks and watches but also kitchen timers, alarm clocks, and
the locks on bank vaults. This has been a revolution in timekeeping. For
hundreds of years we relied on precision engineers to make the machines
that would tell us the time. Now, however, it is all down to physicists.
Mechanical analogue clocks are still made for display or for nostalgia, and
many old clocks still run and keep good time: I inherited a longcase clock
and a Waterbury American clock from my father, and they remain prized
possessions. When I want to know the precise time, however, I look at my
quartz watch, the display on my computer, or the radio-controlled quartz
clock on the kitchen wall.

ABOVE: A desk clock from
1967, when the new quartz
technology began to enter
every home and workplace.

ABOVE: A modern quartz
watch, which will remain
accurate to within a minute
over the course of a year.

A quartz oscillator
vibrates
with a frequency of 32,768Hz
(cycles per second)

In 1967, graduate student Jocelyn Bell helped build a radio telescope at Cambridge University in England to look for the mysterious newly discovered quasars (quasi-stellar radio sources).

Little Green Men?

THEN ON 28 NOVEMBER, Bell noticed an odd piece of "scruff" on her chart paper, which she assumed had been caused by some local source, such as sparking in a malfunctioning car. When it appeared again a few days later she was able to work out that it was not local, but was coming from outer space. Eventually she could predict when it was going to come, and was ready to pick up the trace on high-speed chart paper and therefore in detail.

The result was shocking: a burst of radio waves coming every 1.3 seconds with extraordinary precision. It was coming from a single position around 200 light years away, which means within our galaxy, the Milky Way, but far outside our solar system. Surely it must be a deliberate signal from something? Bell and her co-researchers called it LGM for Little Green Men, which was how aliens were often depicted in science fiction. There was a meeting in the lab: could they possibly publish this, and risk being laughed at by the world? Then, one night just before Christmas, Bell noticed a different bit of scruff, and by rushing out 10km (6 miles) to the telescope in the middle of the night she managed to get a detailed trace: it was a different signal, coming every 1.25 seconds, from another part of the sky.

There could not be two lots of Little Green Men signalling simultaneously, and eventually the sources were found to be neutron stars emitting pulses of radio waves. They were called pulsars (short for pulsating stars), and Bell found the first four; more than 2000 are known today. Bell's supervisor shared a Nobel Prize for the discovery, but she was excluded.

One remarkable thing about pulsars is the regularity of their pulses. Bell's first pulsar, now called by the unglamorous name PSR1919+21, has a period of 1.3373s,

JOCELYN BELL

British astrophysicist Jocelyn Bell was born in 1943 in Belfast, Northern Ireland, and was among the first girls at her school to be allowed to study science. When she first discovered pulsars as a postgraduate student, her supervisor Antony Hewish was highly sceptical of the results. Only Bell's persistence prevented the discovery being dismissed as interference. But she is philosophical about missing out on the Nobel Prize: "I am in good company, am I not!"

and each pulse lasts for just 16ms. The high-speed pulsar PSR B1937+21 has a period that has been measured with a precision of one part in a million billion: 0.00155780644887275ms. For a time, pulsars were considered as the basis for a time standard, since they rivalled the new atomic clocks in regularity, but atomic clocks are more convenient, and have become even more precise.

RIGHT: A depiction of a pulsar, a neutron star, emitting massive bursts of radiation at regular intervals.

BELOW: The plaque on the side of the *Pioneer* spacecraft contains messages for extraterrestrials, including a depiction of the Sun's position relative to the centre of the galaxy and 14 pulsars (left).

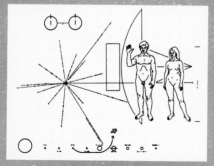

Atomic clocks are the most precise timekeepers yet discovered. The best of them would scarcely lose a second in the whole lifetime of the Universe. And they have revolutionized the very way we perceive time.

ATOMIC CLOCKS

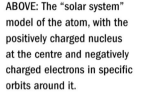

ABOVE: The "solar system" model of the atom, with the positively charged nucleus at the centre and negatively charged electrons in specific orbits around it.

ATOMS HAVE A SMALL NUCLEUS OF PROTONS AND NEUTRONS in the centre surrounded by electrons whizzing around in discrete energy levels – rather like books on a bookcase, with each pair of electrons having their own distinct shelf. The highest energy electrons are kept at the top (away from the nucleus) and the lowest at the bottom (nearest the floor or the nucleus).

Electrons occasionally change levels. There is a fixed energy difference between one level and the next, so when an electron falls down one level it emits a fixed packet of energy, in the form of microwave radiation. Provided the atom is in a stable configuration, this microwave radiation is extraordinarily precise, and this is the basis of an atomic clock. There is no dangerous radiation; no radioactivity; the atom is not disintegrating by nuclear fission. The process is quite harmless.

Harnessing the atom for timekeeping

Lord Kelvin suggested the use of atomic transitions to measure time in 1879, but the necessary technology did not become available until the 1940s. The first atomic clock was an ammonia maser (a maser emits coherent, amplified microwaves) built at the US National Bureau of Standards (now NIST) in 1949. It was primitive, and

THE NEW SECOND

The National Physical Laboratory's 1955 atomic clock led to a new definition of the second. Before that date, one second was defined by the Sun, which appears to go round the Earth in 24 hours, each of which has 60 minutes, each with 60 seconds; so there are 24x60x60 = 86,400 seconds in a day. Because the days are not all the same length, the average is called the "mean solar day", and the second was defined as 1/86,400 of the mean solar day. Since 1955, the second is defined as the duration of 9,192,631,770 periods of the radiation corresponding to the transition between the two hyperfine levels of the ground state of the caesium-133 atom (right). The reason for the change is that the transitions in caesium atoms are much more regular than the Earth's rotation, and scientists especially need precisely defined units of time.

less precise than a quartz clock. The clock built at the National Physical Laboratory in England in 1955 used caesium rather than ammonium. It was found to be precise to one part in ten billion.

Atomic clocks are also made using other elements, including hydrogen and rubidium. Rubidium clocks in particular are relatively cheap and are longer-lasting than the caesium standards. In the early 2000s, physicists at NIST built quantum logic clocks based on single ions (charged atoms) of mercury and aluminium. They claim these are precise to within one second in a billion years, which makes them the most precise clocks yet made. In the future, small, commercially available atomic clocks may become consumer items – you may soon be able to buy one in a wristwatch.

ABOVE: This tiny atomic clock, called a chip-scale atomic clock, is the size of a grain of rice.

Setting the standard

Coordinated Universal Time (UTC; see p159) is based on an average of 260 atomic clocks in 49 places around the world. US standard time is based on the US Naval Observatory and Hydrographic Office (USNO) Master Clock, which in turn is based on 33 independent caesium clocks and a dozen hydrogen maser clocks.

The radio-controlled clocks we use at home are not themselves atomic clocks, but the time signals they receive by radio originate from atomic clocks. All good radio clocks should be consistently precise to within one second.

After 1955, a "second" was defined by relation to the caesium ATOM, not the ROTATION of the EARTH

BELOW: The first atomic clocks were bulky and expensive, and required very high wattages.

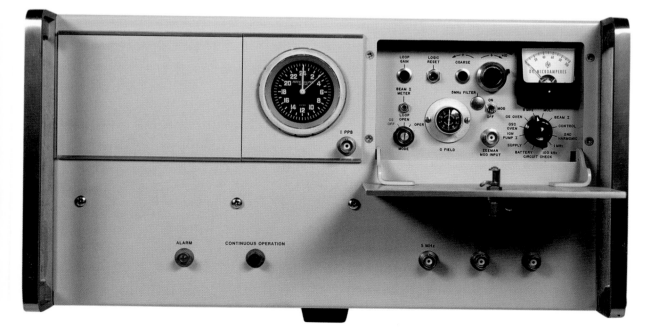

The Global Positioning System (GPS, or officially NAVSTAR-GPS) allows anyone with a suitable receiver to find their location to within about 20m (66ft). It was originally set up for the US military forces in the 1970s, and is still maintained by the US government, but was made freely available to everyone by President Reagan in 1983. There are plans for independent Russian and European systems in the future.

GPS

GPS IS BASED ON ABOUT 30 SATELLITES some 20,000km (12,500 miles) above the Earth's surface, in orbits so arranged that six satellites should always be within line of sight of any point on Earth. Each satellite carries three atomic clocks, and every 30 seconds transmits its position and the exact time. Provided the Earth-based receiver is in contact with at least four satellites, the software on board can work out exactly where it is.

The time transmitted by the satellites is GPS time, which was the same as UTC in 1980, but has drifted about 20 seconds, because GPS time does not include leap seconds. The satellites are travelling so fast that their clocks are affected by special relativity; they run about 38 microseconds slow every day; this has to be corrected. The time is transmitted in the form of weeks and seconds. Week zero started on 6 January 1980 and came round again on 21 August 1999. The seconds go from 0 to 604,799 (the number of seconds in a week) and then start again.

BELOW: A GPS satellite transmits the time and its own position every 30 seconds to Earth-based receivers.

CALCULATING THE GPS RECEIVER'S POSITION

When the receiver gets a signal from a satellite, the signal includes the exact time at which the signal was sent, so the receiver knows how long the signal has been travelling. Since the signal travels at the speed of light, the receiver can work out how far away the satellite was when it sent the signal. It does the same for the next two satellites, and can then work out where it is, by triangulation: a calculation based on knowing your distance from three points. The only problem is that the receiver's clock is unlikely to be precise, so the signal from the fourth satellite is used to correct any error in the receiver's clock.

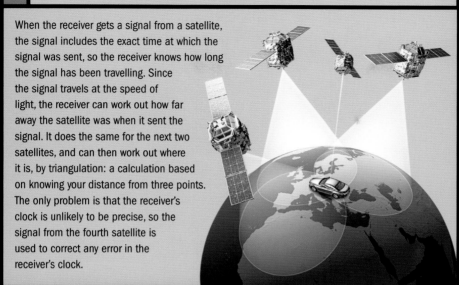

The precise time signals broadcast by GPS satellites are used by banks, mobile phone systems, the controllers of power stations, and many other commercial organizations. GPS is used by ever-increasing numbers of people for navigating in cars or on foot, for surveying and map-making, for emergency services and disaster relief, for recreational activities such as geocaching, and for tracking pets, people, and aircraft. Specially equipped receivers can in principle work out their position to within about 2mm (1/12in), which is immensely useful for surveyors.

Two hundred and fifty years ago, John Harrison showed the Board of Longitude that we can use time to find out where we are. GPS is doing exactly the same thing, but using high technology, and providing extraordinarily accurate results.

BELOW: GPS navigation systems have become commonplace and have entirely changed the way we make journeys.

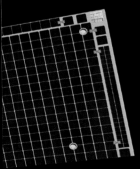

Most people are good at estimating time intervals between a tenth of a second and ten seconds, since this is roughly the range over which the human body functions. Before the invention of pendulum clocks, however, there was no accurate way of measuring seconds.

SECONDS

BELOW: In a sprint race, a false start is called if any of the athletes moves within 100ms of the firing of the gun. This is deemed the minimum amount of time needed for the nerve signals to travel from the ear to the brain and then the brain to the muscles.

ONE SECOND is a fundamental unit in the Système Internationale (SI). It used to be 1/86,400 of the mean solar day, but is now defined as the duration of 9,192,631,770 periods of the radiation corresponding to the transition between the two hyperfine levels of the ground state of the caesium-133 atom. This theoretically applies to a single caesium atom at rest at absolute zero of temperature. Today's atomic clocks are precise to within one second in several million years.

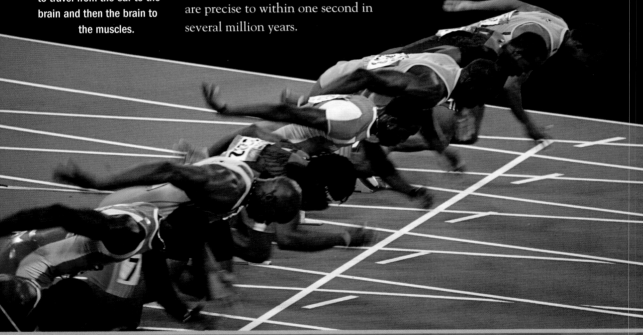

IMAGINING A MILLISECOND

Sound travels just 34.3cm (13.5in) through dry air in one millisecond. Through water, sound travels four times as fast, covering 1.5m (5ft) in a millisecond. By contrast, light travels 300km (186 miles) in 1ms – almost one million times faster than sound in air.

0   5   10   15

Milliseconds

One millisecond (1ms) is one thousandth of a second (10^{-3}s). This is about the time your computer takes to write to or read from the hard drive or a CD. For reflex actions, like pulling your finger from a hot stove, nerve impulses travel at around 100mm/ms, but for more measured thought they amble along at around 10mm/ms. Pain travels even more slowly – at around 1mm/ms. So if you stub your toe on a rock you will feel the pressure in about 20ms, but you won't feel the pain for a couple of seconds. What is even more weird is that if you decide to kick the rock, your nerves will start the muscles moving 500ms before you consciously "make the decision" to do so. When you are mildly stimulated, for example by tickling, you may not become aware of it until it has been going on for 500ms, and then you "backdate" the sensation to match what actually happened. This is "Libet's half-second delay" (see p26).

ABOVE: Some hummingbirds beat their wings 90 times a second when they are hovering. Some flies, on the other hand, can beat their wings more than 1000 times a second.

You can hear two distinct *clicks* if they are at least *2ms* apart

Microseconds

One microsecond (1μs) is 10^{-6}s; a millionth of a second, or a thousandth of a millisecond. Research in cosmology and particle physics has allowed scientists to work out what happened in the first microseconds of the life of the Universe, and they hope by using the Large Hadron Collider to push that frontier back even further, to the first nano- and femtoseconds (see p250). The Indian Ocean earthquake and tsunami of 26 December 2004 shortened our days by about 2.7μs, because it slightly reduced the oblateness of the Earth – made it a little less fat around the middle. The atomic clocks in GPS satellites run slow by 38μs a day because they are moving so quickly.

BELOW: The blink of an eye lasts for around 300ms.

In 1ms, sound travels 34.3cm (13.5 in) in dry air at room temperature... that's a speed of 343m/s (1125 ft/s)

20   25   30   34.3cm

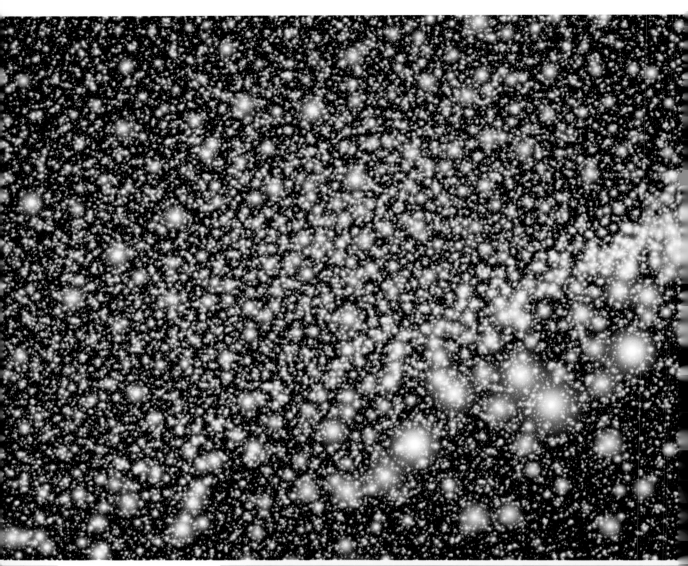

CHAPTER 5

TIME AND SCIENCE

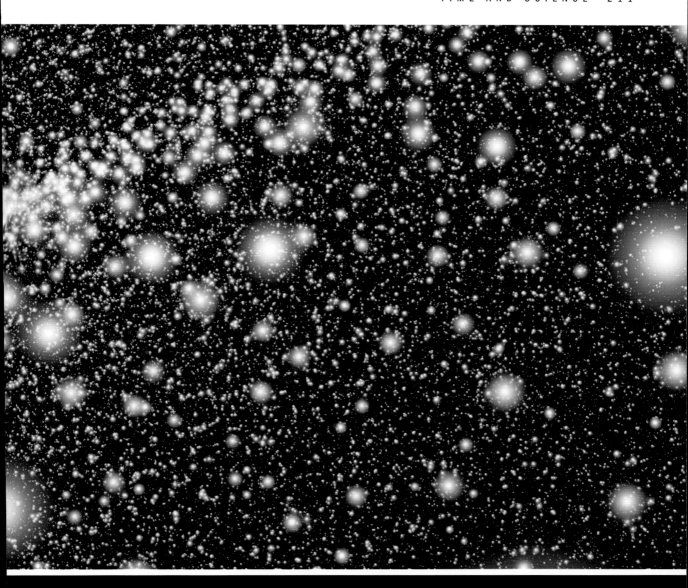

We all use time in our daily lives: our clocks tell us when to get up, what time to eat, and when meetings, trains, and dentists are due. But time is also an integral part of science and scientific research: How fast is a particular chemical reaction? What is the interval between two events? How long do cells take to divide?

Modern science could be said to have started during the Golden Age of Islam, a little over a thousand years ago. Around the year AD 800, Caliph Harun al-Rashid set up in Baghdad the House of Wisdom, where scholars gathered books from Greece and India and translated them into Arabic.

MODERN SCIENCE

BELOW: This painting by English artist William Blake, from 1795, depicts Isaac Newton engrossed with his mathematical measurements and seemingly unaware of the beauty around him. Blake was highly critical of the scientific view of the world, which he saw as a denial of beauty and of God.

THESE SCHOLARS, HOWEVER, DO NOT SEEM TO have applied time and mathematics to their science. That idea seems to have started in the 1590s, when Galileo Galilei began rolling balls down slopes – in effect slowing down the process of falling – and showed that the distance they rolled was proportional to the square of the time; in other words, a ball that rolled 1m (3ft 4in) in 1s would roll 4m (13ft) in 2s. Galileo studied pendulums (see p180) and so laid the foundations for accurate time measurement. He also measured the speed of sound, and tried to measure the speed of light (see p220). He could not work out the mathematics of motion, however: that had to wait for Newton.

Isaac Newton was born in 1642, the year Galileo died. After a good lunch with historian William Stukeley in 1725, Newton claimed that he had worked out his theory of gravity in his "annus mirabilis" of 1665–6. The notion of gravity came to him, he said, while he sat in the garden in contemplative mood, and was "occasioned by the fall of an apple". The curious thing is that he does not seem to have mentioned the apple before this, and it is possible that he made up the story in order to "prove" that he had worked out all the mathematics before his rival Robert Hooke (who was safely dead by 1725). The clue to this suggestion comes from the comet that appeared late in 1681 and was seen again early in 1682. It was very bright, and it went the "wrong way" around the Sun.

If I have seen further than others, it is by standing on the shoulders of **GIANTS**
Isaac Newton

Orbiting objects

Most of the objects in the solar system revolve around the Sun in the same direction. Suppose you were able to plant a video camera in a space ship say a billion miles (1.6 billion km or 1.5 light-hours) "above" the Sun and set it recording the movements of the planets. If you were able to play back the recording, preferably at very high speed, you would see all the planets orbiting the Sun like marbles rolling round the inside of a big bowl, all going in the same direction – anti-clockwise. Their orbits are not quite circular, but slightly elliptical; in other words they are closer to the Sun on one side of their orbits and further away on the other.

BELOW: All the planets and their moons, including Jupiter and its moon Ganymede, and most comets orbit the Sun in an anti-clockwise direction. Halley's comet orbits the Sun in a clockwise direction.

Like the planets, comets have elliptical orbits, but the ellipse is much more pronounced: they whiz past close to the Sun and then way out into deep space beyond the planets before coming back. Most comets orbit the Sun the same way as the planets, but this particular one was retrograde – it went clockwise around the Sun.

In 1644 French philosopher René Descartes had put forward a theory that the planets were whirled around the Sun by enormous vortices in space. However the 1681–2 comet showed that this could not be true – there was no way it could go against a vortex strong enough to move the planets. Isaac Newton dismissed the vortex roundly in print. Nevertheless, seeing this retrograde comet may have prompted him to think seriously about what moved the heavenly bodies. As it happened Edmond Halley was pondering the paths of comets with his coffee-house friends Christopher Wren and Robert Hooke. They could not work out the mathematics; so Halley went to visit Newton in Cambridge to ask for his help.

AN EXPERIMENT TOO FAR

In his 1620 book *Novum organum,* Francis Bacon advocated hands-on science, as opposed to Aristotle's plan, which had been to find the truth by putting enough clever people together and getting them to argue for long enough. Bacon's ideas led to the flourishing of experimental science in the 17th century.

However, Bacon was to pay with his life for his ideas. One wintry day in April 1626 he had the idea that it might be possible to preserve meat by keeping it cold. He jumped out of his carriage on Highgate Hill in London, bought a chicken, and stuffed it with snow. He caught a chill and died a few days later, in the process inventing the frozen chicken.

COMETS – TIME MARKERS FROM SPACE

For thousands of years comets were regarded as harbingers of doom. For example a comet was seen in 1066 (right), and then England was invaded by the Normans and King Harold was killed.

Comets appear in the sky apparently at random. At first they are just glowing points of light, but as they get closer to the Sun they grow tails, which sweep behind them into space and gradually become longer and longer. They get lost in the brightness as they get close to the Sun, and when they emerge the other side, the coma often points forwards.

The head of a comet is really a dirty snowball – a chunk of ice, dust, and rock whizzing through space. Out in deep space the comet is completely frozen, but as it approaches the Sun, it partially melts, creating the tails of dust and vapour. The pressure of the solar wind extends the coma into an enormous tail, which always points directly away from the Sun.

Newton told Halley that he had worked out the mathematics of cometary orbits some time before, but had lost his notes. He promised to do the sums again. The result was that in 1684 he produced a scientific paper "De motu corporum in gyrum" ("On the motion of bodies in orbit"), and then in 1687 he handed to Halley the manuscript of his masterwork *Principia*.

This difficult work, written in Latin, was arguably the most important science book of all time, and laid the foundation for classical physics for the next 200 years. The book might never have been written, however, without the prompting by Halley and his comet. Even then it might never have been published. The Royal Society turned it down; they had already spent their publication budget on a book about the history of fishes. Luckily Halley understood the importance of the work, and paid for the entire cost of publication.

Meanwhile Halley himself profited from the book. He used Newton's laws of motion to work out how the orbits of comets might be affected by Saturn and Jupiter. He also studied reports of earlier astronomers, and became convinced that comets make repeated appearances, the same comet coming back again and again. In 1705 he wrote: "In the year 1456... a Comet was seen passing Retrograde between the Earth and the Sun... Hence I dare venture to foretell, that it will return again in the year 1758." He was right; it appeared on Christmas Day that year. He had died in 1742, but his prediction proved his point; it was an elegant example of the scientific method.

If a scientist proposes a new theory and as a result can make a prediction that turns out to be correct, that is powerful evidence for the correctness of the theory. It turned out that

BELOW: Newton's masterwork *Principia* laid out his three laws of motion, which have come to be known as "Newtonian physics".

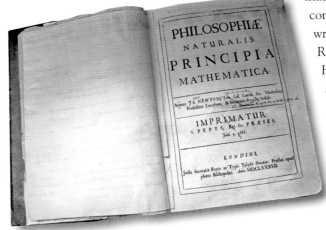

the 1066 comet was the same one, and it had been recorded by Chinese astronomers as long ago as 240 BC. Now it is called Halley's comet, and it will appear again in 2061; make a note in your diary.

Newton's laws of motion were accurate enough to describe and predict the movements not only of planets and comets, but also of Galileo's balls, artillery shells, and even the Apollo missions to the moon 300 years later.

ABOVE: Halley's comet last appeared above Earth in 1986. It will reappear in our night sky in 2061.

Any object **FALLING** near the surface of Earth in a vacuum accelerates by 9.81m/s every second

FALLING OBJECTS

A simple version of one of Newton's equations is

$$s = \tfrac{1}{2}gt^2$$

where s is the distance an object falls in time t, and g is the acceleration due to gravity, which is 9.81m/s/s. It follows immediately when you put the numbers into the equation that an object will fall 4.9m in one second and 19.6m in two seconds.

Strictly speaking, this applies to objects falling near the surface of the Earth in a vacuum, but it remains a good approximation for heavy objects in air. When I dropped tomatoes off the leaning tower of Pisa, 50m (164ft) high, I knew that if I slipped on the polished marble and slid under the rail I should hit the ground after just over three seconds. Plenty of time for a life review.

TIME (S)	1	2	3	4	5
DISTANCE FALLEN(M)	4.9	19.6	44.1	78.5	122.6

Newton's *Principia* was the book that put time and science together, when it became clear that three basic variables are needed to describe objects and events scientifically. These are **mass**, **length**, and **time**. Each has a precise definition.

Fundamental units

MASS IS THE AMOUNT OF MATTER

in an object, measured in kilograms. The world's standard kilogram has for 120 years been a lump of metal (platinum-iridium alloy) made in 1889 and held by the Bureau International des Poids et Mesures (BIPM) at Sèvres in France. A small apple has a mass of around 10g. The Earth has a mass of around 6×10^{24}kg. I have a mass of about 100kg.

I sometimes say that I weigh 100kg, but this is wrong. My mass of 100kg would be the same wherever I was – on Earth, on the Moon, or in deep space. My weight is the force of gravitational attraction between me and a nearby massive object, and should be measured in newtons. A small apple on Earth weighs about 1N (one newton). I weigh about 10,000N on Earth but would weigh only about 1700N on the Moon, and nothing at all if I were in orbit around the Earth.

Length

Length is the shortest distance between two points, and the world standard metre used to be the length of a platinum-iridium bar in Paris. It was originally decreed, under Napoleon's direction, to be one ten-millionth of the distance from the equator to the North Pole, going through Paris. Surveyors took eight years to measure part of this distance, struggling in various war zones, and made a small

BELOW: The ropes on the swing hold all the child's weight. If the father stands behind the child at the bottom of the arc and tries to stop the child, he will be acting against the child's inertia, which arises purely from mass, not weight.

error in the surveying, then a larger error in failing to allow for the flattening of the Earth at the poles. The result was that the first metal bar was 0.2mm short. Another suggestion was that it should be the length of a pendulum that had a period of two seconds, but the force of gravity varies slightly from place to place; so that would not have been a reliable standard.

In 1960 the metre was redefined as the length corresponding to 1,650,763.73 wavelengths (in a vacuum) of the radiation corresponding to the transition between the levels 2p10 and 5d5 of the krypton-86 atom. Once both the passage of time and the speed of light could be measured to high precision, the metre was redefined again as equal to the distance travelled by light in a vacuum in 1/299792458 of a second and this, at least in principle, allows people anywhere in the world to calibrate a standard metre.

So mass is absolute, but our definition of length depends on our measurement of time, and that poor platinum-iridium bar is no longer a world standard; it has lost out to the second and the speed of light.

Time...

...is – well, I tried to explain in chapter one, and I defined the second on p202. Length divided by time gives speed (or velocity, which is speed in a particular direction). So if in one hour you move 2km north, your average speed is 2km/h. Measuring speeds of people, horses, dogs, and so on is relatively simple if they can be persuaded to travel in straight lines for long enough. For example a sprinter who runs 100m in 10 seconds is averaging 10m/s, which is equivalent to just over 22mph.

LEFT: A sprinter who runs 100m in 10s is averaging 10m/s, which is equivalent to just over 22mph.

Measuring things that move very quickly is difficult. A good example is the speed of sound, which was first measured in the early 17th century by French philosopher and scientist Pierre Gassendi. He asked a friend to fire a cannon on a hilltop while he watched from some distance away. He knew that light travels extremely fast; so he must have seen the flash and the smoke very soon after it first appeared, and then heard the bang a bit later.

THE SPEED OF SOUND

ABOVE: Count the seconds between seeing a flash of lightning and hearing the thunder, and that will tell you how far away the lightning struck. A three-second delay means that the lightning is 1km (0.6 miles) away.

THE DELAY IN HEARING THE BANG was therefore due to the travelling time of the sound. His result of over 450m/s was rather too high, possibly because he did not have a good timekeeper, but in the 1650s the Italian physicists Giovanni Alfonso Borelli and Vincenzo Viviani achieved a much better result: 350m/s. This was most impressive given that they cannot have had accurate pendulum clocks for their timing.

During a thunderstorm you may see the lightning flash before you hear the crash of the thunder. Count the number of seconds between the flash and the bang and divide by five (for miles) or three (for km) to find out how far away the lightning strike was. Roughly speaking sound takes 5 seconds to go a mile or 3 seconds to go a kilometre; it takes a millisecond to travel 30cm (1ft).

Sound travels much faster in water (1,484m/s) and in solids such as iron (5,120m/s), but does not travel at all through vacuum, which was demonstrated by Robert Boyle in the 17th century. The speed of sound in water was measured heroically in 1826 by Swiss scientists Jean-Daniel Colladon and Charles-Francois Sturm. They sat in separate boats 16km (10 miles) apart on Lake

MEASURING SOUND

Surprisingly and satisfyingly, you can measure the speed of sound yourself. You need to find a high wall, or an isolated building, or cliff, from which you can get an echo. Stand about 100m (328ft) away and start clapping at half-second intervals. Ideally get a friend to time your claps so that you fit 20 into 10 seconds.

Walk slowly towards the wall, still clapping, until the echo of each clap is exactly in time with the following one. Then you know that the sound of each clap is taking exactly half a second to get to the wall and bounce back to you. Measure your distance to the wall; big strides will be about a metre, and this will give you an approximate measure.

You will probably find that you are between 90m (295ft) and 80m (262ft) from the wall. Suppose it is 85m (279ft) ; then the sound has travelled twice 85m (279ft) or 170m (558ft) in half a second. So the speed of sound is 340m/s. In practice, in dry air at 20°C (68°F), sound travels at 343m/s (1,126ft/s) or 1,236km/h (768mph). This is sometimes known as Mach 1, after Austrian physicist Ernst Mach. The speed of aircraft is sometimes measured in Mach numbers, so that Mach 2 is twice the speed of sound.

Geneva. Sturm rang a bell under water, and at the same time ignited a sample of gunpowder. Colladon watched for the flash and then stuck his head under water (or perhaps he prosaically used an ear trumpet) to listen for the sound of the bell. The result at 15°C (59°F) was about 1,300m/s (4,300ft/s) – not far from today's value.

ABOVE: A jet travelling faster than the speed of sound creates a sound called a "sonic boom". The boom causes water vapour to condense into a conical cloud.

Light travels as fast as anything in the Universe. Radio waves and other forms of electromagnetic radiation go at the same speed, but nothing travels faster than light. The fact that it travels very quickly was obvious hundreds of years ago. Measuring just how fast it does move, however, was a tough challenge.

CHASING LIGHT

GALILEO TRIED MEASURING the speed of light by getting two people to stand on mountain tops with lanterns. The first man opened his shutter to let the light out, and as soon as the second man saw the light he opened his shutter. When the first man saw the light from the second he would know that the light had travelled to the far mountain and back again.

The problem was that light takes only a tiny fraction of a second to travel many miles, and any delay observed by the first man between opening his shutter and seeing the light from the far mountain would be caused almost entirely by the reaction times of the two experimenters. So they were unable to measure the speed of light this way. Galileo concluded that light might travel instantaneously, and that its speed was at any rate at least 10 times faster than that of sound.

The first person to get any sensible measure of the speed of light was Danish astronomer Ole Roemer, who in 1675 was observing the four largest moons of Jupiter, discovered by Galileo 65 years earlier. He knew that as they circled the planet they kept regular schedules – like buses, but much more predictable. What puzzled him was that the orbital timetables of these moons seem to vary through the year, in other words with the position of the Earth in its orbit around the Sun.

When the Earth was further away from Jupiter the moons seemed to get behind schedule; when the Earth was closer they got ahead of schedule. He didn't believe the position of the Earth could really affect the orbits of the moons, and he concluded the difference must lie in the fact that light takes some time to cross the diameter of the Earth's orbit. Using what was then believed to be the size of the Earth's orbit, he deduced that the speed of light was around 200,000km/s.

In the middle of the 19th century a French physicist, Leon Foucault, made a better measurement in his laboratory. He bounced a beam of light from a rapidly rotating mirror to a distant fixed mirror, and then back to the rotating mirror once more. While the light was travelling to the fixed mirror and back the mirror had rotated a little; so the light came off at a slightly different angle. Knowing the length of the light path, the rotation speed of the mirror, and the angle at which the light was reflected back, Foucault was able to calculate the speed of light to be 299,796km/s. This is only about 0.001 per cent greater than today's figure of exactly 299,792.458 km/s.

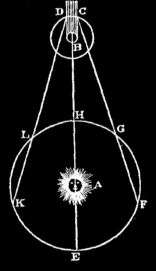

ABOVE: An illustration from the 1676 article on Rømer's measurement of the speed of light. Rømer compared the duration of Io's orbits as Earth moved towards Jupiter (F to G) and as Earth moved away from Jupiter (L to K).

ABOVE: Light travels so quickly that measuring its speed was a major scientific puzzle for hundreds of years.

FOUCAULT'S PENDULUM

Foucault is most famous for his clever demonstration that the Earth is rotating. In 1852 he suspended a heavy iron ball on the end of a 67m (220ft) wire from the top of the dome of the Pantheon in Paris. The pendulum swung to and fro in a stately way; because of the length of the wire, it took slightly more than 16s to swing forwards and back again, and because of its large mass (28kg) it continued to swing for many hours.

What was most intriguing, however, was that the plane in which it moved appeared to rotate clockwise by some 11° per hour. The reason for this was that the pendulum continued to swing in the same celestial plane – in the same direction relative to the stars. Meanwhile the Earth was rotating underneath it; so from the point of view of anyone standing there on what seemed to be a stationary floor, the pendulum was twisting.

This demonstration that the Earth was turning astounded both scientists and public. It continues to fascinate people, and was one of the things that persuaded me to study science. There are many Foucault pendulums around the world, notably at the United Nations in New York, and at the Science Museum in London.

In empty space the speed of light, usually shown by the letter *c*, is exactly 299,792.458km/s – in other words close to 300,000km/s or 3×10^8m/s, which is about 186,000 miles/sec, 1 billion km/h or 671 million mph. To think about it in everyday distances, light travels almost exactly 30cm (1ft) in one nanosecond, a billionth of a second.

The speed of light

ABOVE: Light takes 1.3 seconds to reach us from the Moon, and 8 minutes to reach us from the Sun.

LIGHT TAKES ABOUT 1.3s TO REACH US from the Moon, 8 minutes from the Sun, and 4 years from the next nearest star, Proxima Centauri. Light travels more slowly through glass and other materials that are more dense than a vacuum. Through air it travels very close to its maximum speed.

In normal life, we are rarely conscious of the speed of light, because it is so fast. For example a message can travel around the world in about 140ms, or less than one 7th of a second. If you are making a phone call to someone on the other side of the world and the message goes by optical fibre, it could take as long as a tenth of a second (100ms), since the fibre may not take shortest course, and light travels more slowly through glass.

On the other hand if you speak to somebody via a satellite then there will be a much longer delay. Communication satellites are usually in orbits 35,400km (22,000 miles) above the Earth, so a signal going to a satellite and back takes about a quarter of a second, and you have to wait half a second for a response, which is just noticeable during speech. When Neil Armstrong and Buzz Aldrin landed on the Moon in 1969, they received a telephone call from President Nixon; it was awkward, because the radio waves took about 1.3 seconds to reach the Moon, which meant that when Nixon asked a question he had to wait 2.6 seconds for a response.

Towards the end of the 19th century the speed of light began to show up in unexpected places, and scientists realized that it was fundamental to a whole lot of the physics that was being discovered at the time. One famous experiment was done in 1887 by Albert Michelson and Edward Morley in Cleveland Ohio.

The most famous failed experiment

In the 17th century Robert Boyle had investigated the properties of a vacuum, and had discovered that sound will not travel through a vacuum, but light will. Reasoning that air is necessary to carry sound waves, he and other scientists assumed that there must be some medium to carry the light waves. They called this medium the "luminiferous aether", or sometimes just the "ether", which permeated the Universe, and carried the light to us from the Sun and the stars.

Michelson and Morley reasoned that if the Earth was hurtling through the ether as it rotated and orbited around the Sun at some 30km/s (18.6 miles/s), then the speed of light would vary according to whether it was travelling in the same direction as the Earth or at right angles. They therefore set up an elaborate version of Foucault's experiment, and measured the speed of light with enormous precision, in two directions at right angles. They expected that the beam of light that had travelled up and down the direction of the "ether wind" would take longer than the beam of light which had travelled across it, because it would be greatly slowed down while travelling "upstream". This is like riding a bicycle; you achieve a lower average speed going up and down hills than on the flat, because you lose so much time going up each hill, and you don't make up as much time going down the next one. Indeed the speed of light should be slightly affected, they thought, according to whether the light was moving in the same direction as the Earth, or at right angles to it.

The result was complete failure. To within experimental error they measured no difference at all in the speed of light in the two directions. The speed of light was the same in all directions. In effect they had disproved the existence of the ether.

Michelson and Morley had also shown that the velocity of light remains constant under varying conditions. This was interesting, because of the equations dreamed up by the Scottish physicist James Clerk Maxwell in 1864. Maxwell was a brilliant scientist who among other things took the world's first colour photograph. He collected all the various theories and pieces of evidence on electricity and magnetism from Michael Faraday and other scientists and put them all together into a mathematical theory of electromagnetism.

Maxwell's equations showed that light and any other electromagnetic radiation must travel at the same speed, regardless of how fast the source was moving, and in which direction. One consequence of this was that Maxwell realized that light was just one example of electromagnetic radiation; others are gamma rays, x-rays, ultraviolet and infrared radiation, microwaves, and radio waves.

BELOW: The spectrum of visible light is just one small part of the whole electromagnetic spectrum.

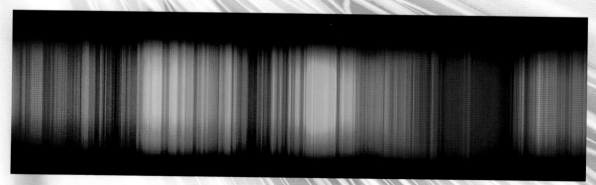

The basic concept of relative motion is straightforward. I am writing this on a train, travelling in a straight line on a level track at a steady 22m/s, or 50mph. Each coach has seats on either side, with a passageway up the middle. If I get up and walk slowly towards the front of the train I shall be moving at say 1m/s (2mph) as seen by someone in one of the seats.

RELATIVE MOTION

HOWEVER, LET US IMAGINE THAT MY ALTER EGO (AE) is standing on the platform as we go through a station. From his point of view I shall be travelling at 23m/s (52mph), which is the combination of the speed of the train and my speed in the same direction in the train.

If I then turn round and walk back to my seat at the same speed relative to the train, my speed relative to AE will be 21m/s (48mph).

This idea of relative motion can be used to derive Newton's first law of motion, which states that an object on which no external force operates either remains stationary or continues to move in a straight line with a constant speed. A football sitting on a horizontal table and neither kicked nor pushed will stay where it is. That is the first part of Newton's law. Now suppose that football is on the train with me, moving at a steady 22m/s. No pushing; no kicking. Again, it stays put from my viewpoint; but from AE's viewpoint the football is travelling at a constant 22m/s in a straight line.

RIGHT: With no external force acting upon her, a train passenger sat in her seat is entitled by the laws of physics to consider herself to be stationary. However, to an observer standing on the platform, she is whizzing past at a speed of many kilometres per hour.

Looking at relative speeds can help to solve some tricky problems. Imagine two balls of the same size, both perfectly elastic and bouncy, but one very light and the other much heavier – for example a ping-pong ball and one of those super-bouncy hard rubber balls, sometimes called a superball. Hold them with the ping-pong ball on top of and touching the superball, and drop the two together from a height of say 1m on to a hard surface, such as a stone floor or a hard worktop. Dropping them together is quite tricky, but if you get it right then when they hit the surface the superball will bounce to near your hand, but the ping-pong ball will bounce several metres into the air, and should hit the ceiling of any normal room. This is an impressive sight, and I never understood it until I read David Mermin's book *It's about time.*

The explanation comes in six steps:

1. Imagine the two balls on a hard table, where they can roll without friction. The ping-pong ball rolls towards the superball at a speed of 2m/s. When it collides, the superball barely moves, because of its large mass, but the ping-pong ball rebounds with speed 2m/s. In other words it behaves as though it had hit a brick wall, bouncing off with the same speed, but in the opposite direction.

2. Now imagine that the ping-pong ball is stationary, and the superball rolls into it in an easterly direction at speed 2m/s. The superball will carry on in the same easterly direction with almost the same speed, but what about the ping-pong ball?

3. We can answer this question by using artificial relative motion. Put the table on a train which is travelling west at 2m/s. Now from the point of view of an observer on the ground (AE) the superball is stationary, while the ping-pong ball is rolling westwards towards it

4m/s

12m/s

4m/s

LEFT: After being dropped from a height of 1m, both balls are travelling at 4m/s just before they hit the ground (above). After hitting the ground, the superball rebounds at 4m/s, but the ping-pong ball rebounds at 12m/s (below). The superball bounces back up to about 1m, but the ping-pong ball bounces 9m into the air.

DOES OXFORD STOP?

To Einstein all motion was relative; so movement of the countryside was equivalent to movement of the train. When he travelled to Oxford University to give a lecture, he got on the train at Reading, and allegedly asked the train manager "Does Oxford stop at this train?".

at 2m/s. We know what happens then, from 1 above: the ping-pong ball bounces off towards the east at 2m/s. But the train is moving west at a speed of 2m/s; so from my point of view in the train the ping-pong ball moves east at 4m/s. In other words the ping-pong ball gets batted off at twice its own speed. This would be the same if the train were stationary; the superball moving at 2m/s would bat the stationary ping-pong ball off at 4m/s.

4. Now imagine the balls rolling towards each other, each moving at 2m/s, and meeting head-on. The superball will blunder on, but the ping-pong ball will bounce back at 4m/s as a result of being bashed by the superball (from 3 above) plus 2m/s from the reversal of its own motion; so it will bounce off at 6m/s – three times its original speed.

5. Finally think about the two balls dropped on to the floor. When dropped from a height of 1m, the superball will hit the floor at about 4m/s; then it will bounce upwards at the same speed, 4m/s. At this instant the ping-pong ball is still moving down at 4m/s; it meets the superball coming upwards, and the result is like the previous paragraph. The superball bats the ping-pong ball off at twice its own

BELOW: Einstein first explained his revolutionary theory of special relativity in terms of observers standing in or next to a moving train. His simply stated explanation remains the best way to grasp the subtleties of his difficult idea.

speed – ie 8m/s, and in addition the ping-pong ball will bounce from the collision with another 4m/s of its own making. The ping-pong ball therefore flies up at three times its original speed, or 12m/s.

6. The height to which a ball bounces is proportional to the square of its upwards speed; so in principle, if there were no friction or air resistance the ping-pong ball would bounce to nine times as high as the place it was dropped from, in other words 9m. It should hit the ceiling.

The speed of light

Relative motion becomes tricky when things are approaching the speed of light. Suppose my train is actually a supertrain (unlikely in the UK, but we can imagine) which gets faster and faster until it is moving at about two thirds of the speed of light, or 200,000km/s. At this point I throw a ball forward down the train with such force that it leaves my hand at the same speed, 200,000km/s. Relative to AE, therefore, you might expect that the ball would be travelling at 400,000km/s. But nothing can go faster than light; the Universe has a speed limit of 299,792.458km/s.

LIGHT TRAVELS at 186,000 miles per second, or 671 million mph

The *speed of light* never varies. Ever

This was the sort of conundrum that faced Albert Einstein in 1905 when he was thinking of writing a paper on what came to be called special relativity. He started with two postulates: first that there is no such thing as a frame at rest, because without acceleration you can't tell whether or not you are moving. From my seat in the train, I know I am moving because of all the bumps in the track, but apart from that I could be sitting still, while the landscape flashing past could simply be on a huge conveyor belt. Likewise on the surface of the Earth we do not have a sense of movement round the Sun or round the galaxy; as far as our senses go we are at rest. Einstein's second postulate was that the speed of light is always the same.

The relativistic velocity addition law

This law states that if an object is travelling at speed v_1 in the frame of a train that is moving in the same direction at speed v_2, then the speed of the object from the point of view of a person standing beside the track will be

$$(v_1 + v_2)/(1 + (v_1 v_2/c^2))$$

where c is the speed of light. When v_1 and v_2 are small compared to c, which is the case for almost everything in real life, then this is almost identical to the "normal" value $(v_1 + v_2)$. When either v_1 or v_2 approaches the speed of light then the combined speed is significantly reduced.

BELOW: AE compares stationary time with moving time. Which is slower?

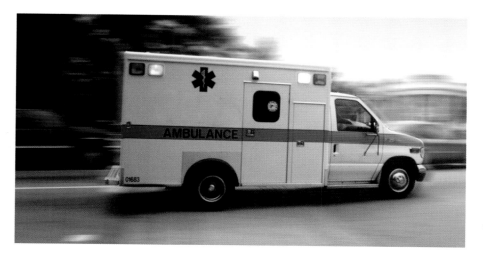

LEFT: The Doppler effect causes an ambulance siren to go down in pitch as it passes you in the street because the sound waves bunch in the direction of travel. The same thing happens with light waves.

Suppose while I am sitting in my train, proceeding at a sedate 22m/s, I take a photograph of the person sitting opposite me. The flashgun in my camera sends out a pulse of light at the speed of light, 299,792,458 m/s, or c for short. This is the speed of the pulse of light relative to the train. (Note this is the speed of light in a vacuum. In air, light travels a little more slowly, but I shall ignore that.)

You might expect the speed of that pulse relative to someone watching from the ground to be $(c+22)$m/s, since when I walked along the train, I was able to add my walking speed to the speed of the train. Instead it turns out that the pulse of light travels at exactly c as seen from the ground. In other words, light always travels at c, regardless of the speed of the light source, as predicted by Maxwell.

We should not be surprised by this. The speed of sound does not vary with the speed of the source. When an emergency vehicle passes you in the street you hear the note of its siren drop as it goes past, because the sound waves are bunched up as it approaches you – the wavelength decreases – and are stretched out as it recedes – the wavelength increases. This is the Doppler effect. The speed of the sound, however, does not change. The same thing happens with light waves. When the source moves towards you, the wavelength of the light decreases, causing a blue shift; when it recedes the wavelength increases, causing a red shift (see p246). The speed of light remains the same.

BELOW: If a light source is travelling towards you, its light waves shorten, causing blue shift. If it is travelling away from you, its waves lengthen, causing red shift.

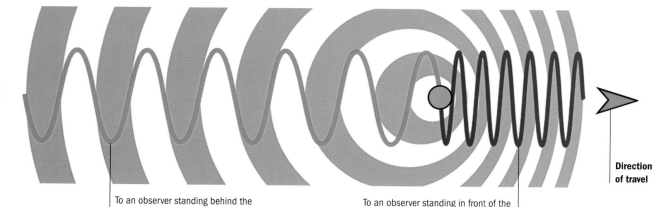

Direction of travel

To an observer standing behind the light source, the light waves lengthen.

To an observer standing in front of the light source, the light waves shorten.

A moving train does cause some conundrums. As a photographer, I sometimes use more than one flashgun to light my subject, one flashgun on the camera and one or two "slave" flashguns, which fire as soon as they detect the primary flash.

SPECIAL RELATIVITY

BELOW: The light from a camera's flash travels towards you at the same speed – c – regardless of your speed relative to the camera.

SUPPOSE I AM IN THE MIDDLE OF MY RAILWAY COACH, travelling at 22m/s, and I have put one slave flashgun at each end of the coach. When I fire a flash from the camera the light will travel both ways down the carriage, and both slaves will fire simultaneously. At least, the flashes will fire simultaneously from my point of view, but not for someone like AE standing by the track.

For AE it is clear that the back of the coach is moving towards the camera, closing the distance between them while the pulse of light is travelling; the light is still travelling at c, but the distance to the back of the coach is decreasing while the light is in transit. Therefore the pulse of light going backwards has less far to travel than the pulse going forwards, because the front of the coach is moving away from the camera. It follows that the rear flashgun will fire before the front flashgun. So AE would disagree with me, and assert that the slave flashes were not simultaneous.

This is one of the consequences of what came to be called "special relativity", explained in Einstein's paper of 1905. Whether or not two events in different places are simultaneous depends on your point of view, or, more technically, on your "frame of reference". Einstein's first postulate was that there is no absolute frame of reference; so both of these views of the timing of the slave flashes are equally valid.

What I see as simultaneous on the train may not be simultaneous from a different "frame of reference". In Newton's Universe, time was the same for everyone – indeed he wrote "Absolute, true, and mathematical time, of itself, and from its own nature, flows equably without relation to anything external." In Einstein's Universe time is not fixed. Instead it varies according to your point of view.

Suppose that I use my light flashes to synchronize clocks at each end of my coach. I know the clocks are in sync because I am in the middle of the coach, and the speed of light is constant; so it must have taken the same time to reach the two ends. To AE, however, the front clock trails behind the rear clock by an amount that turns out, while we are travelling at 22m/s, to be 4.9×10^{-15}s, or 4.9fs (femtoseconds). This may be an exceedingly short period of time, but it is real. So from AE's point of view my front clock is running behind.

Imagine further that below each slave flashgun there is a slit across the floor of the coach, so that some light from the slave passes through and burns a mark in a piece of light-sensitive paper between the rails. How far apart will

be 20m apart, because the slave flashes fired simultaneously and the coach
is 20m long.

From AE's viewpoint, however, the marks must be further apart than the length
of the coach, because the rear flash fired first, and the train must have travelled some
extra distance before the front flash fired. From AE's viewpoint, therefore, the train
is shorter than I perceive it to be. In fact it is shorter in the same proportion as the
front flash is later than the rear one. In other words, to outside observers, moving
objects shrink in the direction of travel.

Suppose my supertrain is going at 200,000km/h or 2×10^8m/s then the rear flash
will fire $(20 \times 2 \times 10^8)/(3 \times 10^8)^2$ seconds before the front; that is 4.4×10^{-8}s or 44ns.
During this time the train will have moved 8.9m. Meanwhile from AE's viewpoint
the train shrinks by $\sqrt{(1-v^2/c^2)}$ which is $\sqrt{(1-4/9)} = 0.75$; so it becomes 15m long.

TRAINS AND FLASHES

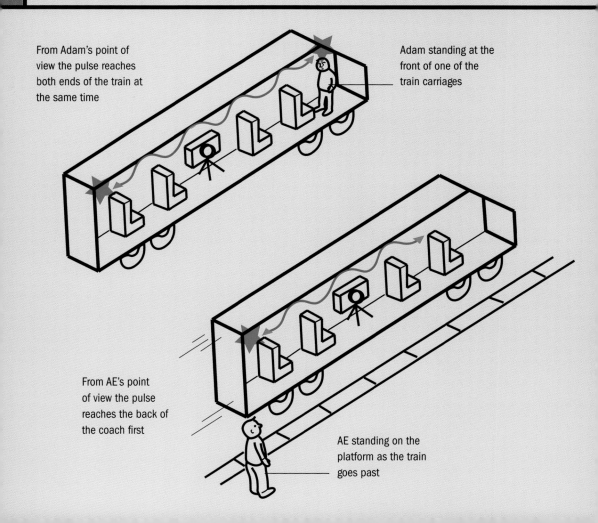

From Adam's point of
view the pulse reaches
both ends of the train at
the same time

Adam standing at the
front of one of the
train carriages

From AE's point
of view the pulse
reaches the back of
the coach first

AE standing on the
platform as the train
goes past

Albert Einstein was born in the German city of Ulm in the year in which James Clerk Maxwell died, but the family moved to Munich, and in 1894 to Italy. Albert stayed at school in Munich and went on to study in Switzerland, where he trained as a teacher of physics and mathematics. In 1901, he received his diploma, but he could not obtain a teaching post, and so he took a job as assistant examiner at the patent office in Bern.

BELOW:
Albert Einstein, photographed in 1921, the year he was awarded the Nobel Prize for Physics.

Albert Einstein (1879–1955)

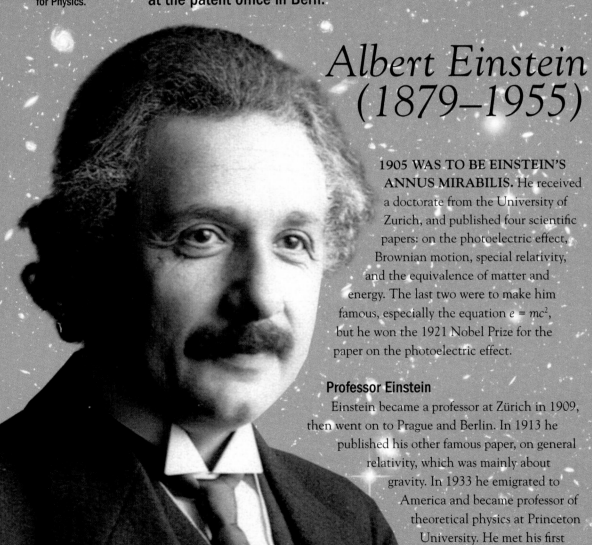

1905 WAS TO BE EINSTEIN'S ANNUS MIRABILIS. He received a doctorate from the University of Zurich, and published four scientific papers: on the photoelectric effect, Brownian motion, special relativity, and the equivalence of matter and energy. The last two were to make him famous, especially the equation $e = mc^2$, but he won the 1921 Nobel Prize for the paper on the photoelectric effect.

Professor Einstein

Einstein became a professor at Zürich in 1909, then went on to Prague and Berlin. In 1913 he published his other famous paper, on general relativity, which was mainly about gravity. In 1933 he emigrated to America and became professor of theoretical physics at Princeton University. He met his first wife Mileva Maric while they were both training to be

THE LORENTZ FACTOR

Although he was a brilliant theoretical physicist, Albert Einstein struggled with mathematics, and he relied on others to provide the mathematical tools with which to describe his theories of relativity, including equations formulated by the Dutch physicist Hendrik Lorentz (1853–1928).

The "Lorentz transformation" describes mathematically how the measurements of time and space made by an observer in one frame of reference can be converted into the measurements of a second observer in a different frame of reference. Lorentz's equations showed that the effects of special relativity changed the results of Newtonian physics when moving from one frame of reference to another by a factor known as the "Lorentz factor". Where velocities are low, the Lorentz factor equals nearly 1, so Newton's equations give good results. Only at velocities that are a significant proportion of the speed of light does the Lorentz factor reveal large errors in the Newtonian view of the Universe.

teachers at the Polytechnic in Zurich. They had a daughter Lieserl in 1902, married in 1903, and had two sons, Hans Albert and Eduard. Albert and Mileva divorced in 1919, and Einstein then married his first cousin Elsa Lowenthal, who died in 1936.

The political scientist

In later life, Einstein became much involved in politics. At the beginning of World War II he wrote personally to the President of the United States, Franklin Roosevelt, to warn of the danger that Germany might develop an atomic bomb, and suggested that the Americans should build one themselves. He later said that he regretted his role in the development of the atomic bomb. He was offered the presidency of the state of Israel, which he declined, and he worked with Chaim Weizmann to establish the Hebrew University in Jerusalem.

At the age of four Einstein had been fascinated by the compass that his father showed him, and in his late teens he became intrigued by the work of James Clerk Maxwell. He imagined riding alongside a beam of light, and wondered what it would look like. It should look like a stationary wave, but Maxwell's own equations had shown this was impossible, and Einstein wondered what was wrong with his idea. That was one of the notions that led him to the constancy of the speed of light, and special relativity.

Einstein's theories of relativity remained theoretical for some years after they were published, but the predictions they made have now been shown to be correct. The general theory of relativity provides the most satisfactory account of gravity.

These relativistic ideas are bizarre, and they affect time. Moving clocks tick more slowly than stationary clocks. When AE on the ground watched my flashes going off, he would say that the front one went off later than the rear one, and would probably say that if I think they were simultaneous that must be because, assuming I had clocks at both ends of the coach, my front clock was running slowly. And indeed from his point of view both my clocks were really running slowly.

SLOW CLOCKS

THE ONLY THING WE KNOW TO BE CONSTANT is the speed of light; so let's base a clock on that. I put one of my slave flash guns on one side of the carriage and fire the camera flash from the other side 3m away, with the other slave beside it. The slaves then fire one after the other as the bursts of light cross the carriage to and fro. The first will fire 10ns after the camera, since light travels 30cm in 1ns and therefore 3m in 10ns. The second will fire 10ns later, and so on. They form a clock, which can tick forward 1s after 100 million flashes. (In practice, real flashguns will not charge up so quickly, but we can imagine…)

From the track, however, the situation looks rather different. Because the train is moving, AE sees that each burst of light has to travel diagonally across the coach to reach the other flashgun, and therefore has to go more than 3m (10ft). Therefore it must take longer than 10ns to get there; so if my clock says it gets there in 10ns then my clock is running slow.

THE HAFELE-KEATING EXPERIMENT

In October 1971 Joseph Hafele and Richard Keating took four caesium atomic clocks round the world on commercial airliners. When they returned they compared the time on their clocks with those at the United States Naval Observatory. After one circuit flying eastwards the clocks were about 60ns slow, as predicted. There were further complications from gravitational effects, but the results were just what they expected from special relativity. Further experiments in 1976 and in 1996 confirmed these results with a much higher degree of precision.

The fact that time was different for Hafele and Keating than for the scientists at USNO is simple proof that Newton was wrong. Time is not absolute and universal: it is elastic, and varies from person to person.

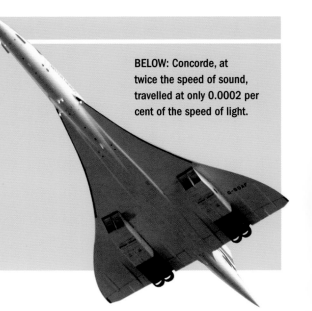

BELOW: Concorde, at twice the speed of sound, travelled at only 0.0002 per cent of the speed of light.

Accelerating objects become heavier

Einstein's formula for the factor by which clocks are slowed is $\sqrt{(1 - v^2/c^2)}$ – the "Lorentz factor" (see p233) – where v is the speed of the clock and c the speed of light. The effect is small at normal speeds: even at half the speed of light a clock would run only 13 per cent slower than normal; but at 99 per cent of the speed of light it would run at only one seventh of its normal speed.

This sounds like a piece of airy-fairy theory, but it is firmly backed up by evidence. First, atomic clocks have been flown around the world and shown to be slow when they returned to base (see below left).

MOVING ATOMIC CLOCKS

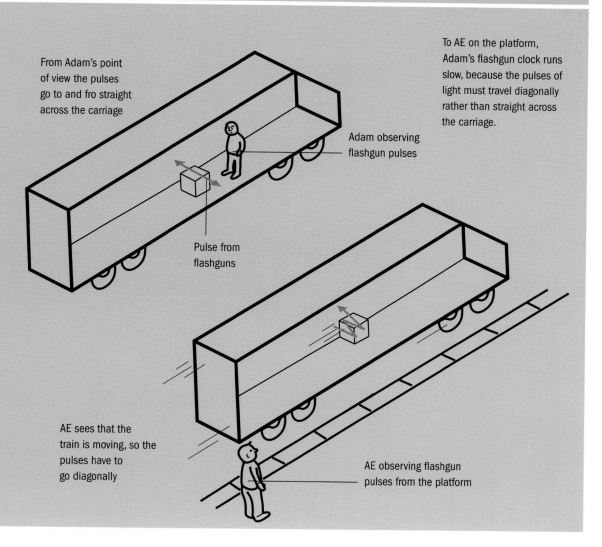

From Adam's point of view the pulses go to and fro straight across the carriage

To AE on the platform, Adam's flashgun clock runs slow, because the pulses of light must travel diagonally rather than straight across the carriage.

Adam observing flashgun pulses

Pulse from flashguns

AE sees that the train is moving, so the pulses have to go diagonally

AE observing flashgun pulses from the platform

Second, the atomic clocks in GPS satellites (see p206) are travelling at considerable speed – about 8,000 m/s or 18,000 mph. This means that as seen from the Earth those clocks are running slowly, by about 38μs (microseconds) per day. In order to have precise time correlation, which is necessary for accurate positioning, a relativistic correction has to be made, by on-board electronics, to the time the signal which they send out.

Third, there are fundamental particles (fragments of atoms) called μ-mesons (mu-mesons), which are formed when cosmic rays come hurtling in from the Sun

BELOW: A computer image of particles colliding in the Large Hadron Collider (bottom). They travel at almost the speed of light, but they can never be made to move at the speed of light.

and collide with atoms in the atmosphere. They normally survive for only about 2μs (2 microseconds), which would allow them to travel about 1,000m (3,280ft) downwards. In other words they should all decay in the upper levels of the atmosphere. In practice about half of them reach the surface of the Earth. This is because they are travelling so fast that their "clocks" are running enormously slowly, so that by our time they last not 2μs but more like 100μs.

Aside from the fact that moving clocks tick more slowly, other strange features of Einstein's Universe are that moving objects shrink and get more massive, by the same factor, which is $\sqrt{(1-v^2/c^2)}$. This means that

Astronauts travelling at 99 per cent of the speed of light will age just one year in a seven-year trip

in my train trundling along at 22m/s everything has become shorter and more massive by about 2.5×10^{-13} per cent; in other words from AE's point of view my coach has shortened by 50fm from 20m to 19.99999999999995m, and I am one quarter of a nanogram heavier, which luckily is not enough to register on the bathroom scales.

For the particles hurtling round the racetrack of the Large Hadron Collider, however, the change in mass is all too real. They are accelerated to 99.9999 per cent of the speed of light, which means that their mass is increased by a considerable factor. In fact as the particles go round and round, and more energy is pumped into the system, the main change is that they get not faster but more and more massive, by a factor of 100 or more, so that they carry much more energy into any collision that happens.

They can never reach the speed of light, however, because they would become infinitely massive and to accelerate them that last little bit would need an infinite amount of energy.

One potential consequence of the relativistic slowing of clocks is that astronauts will age less while they are travelling than when they are Earthbound. If a spaceship could be made to travel at 99 per cent of the speed of light, then all clocks on board, including biological clocks, would run about seven times more slowly than usual (as viewed from Earth). Therefore if they travelled for seven years (to get to a nearby star) they would have aged by only one year. This raises the possibility of time travel into the future.

ABOVE: On a spaceship travelling at 99 per cent the speed of light, its clocks would tick seven times more slowly than a clock on Earth, if measured by an observer on Earth.

OPTICAL STRETCHING AND SHRINKING

There is another subtlety here. If you watch a train (or other object) coming towards you, you will actually see it as longer than it really is. Light coming from the far end of the train must have taken a few nanoseconds longer to reach you than light from the near end, which means that the light from the back end must have left a few nanoseconds earlier than light from the front. So if the train is 100m (328ft) long and travelling at 22m/s (50mph) you will see it stretched by some 7µm, or about one tenth of the thickness of a human hair.

As the train goes away from you it will appear to shrink by the same amount. Note that these effects are not real; they are merely visual illusions – tricks of the light. The relativistic effects, however, improbable as they sound, are real: the train really does shrink. A rocket travelling towards you at nearly light speed really does, from your frame of reference, become squashed.

Time travel has been a feature of many stories, books, and films, starting with the famous tale *The Time Machine* written by HG Wells in 1895. Wells's time traveller described time as the fourth dimension, anticipating Einstein by ten years, and built a machine that whisked him into the future. Since then countless fictional machines have whisked countless time travellers into the future, sometimes using methods on the edge of known physics and engineering, sometimes just pure fantasy. But is time travel really possible?

TIME TRAVEL

WE CAN TRAVEL FROM NEW YORK TO LOS ANGELES because both places are there now, but can we really travel to the future, which does not yet exist? At least, it doesn't exist for ordinary people, but it does for Einstein and his physicist followers. Since time depends on who is measuring it, and where, how can we assert that the future has not arrived yet? Or to put it another way, that "now" which philosophers have argued about for centuries has no absolute meaning.

Yes, in theory, time travel is possible. Suppose a team of astronauts was sent off in a spaceship to Sirius, the dog star, the brightest star in the night sky. They set off from Earth with a steady acceleration of 2g, so that they would feel twice their normal weight, but this should not be too uncomfortable. Sirius is about 8.6 light years away; so that by the time they got half way there they would have been travelling for about two years, and would have reached around 99 per cent of the speed of light. At this point they would have to turn their ship around and start decelerating in order not to overshoot Sirius.

When they get to Sirius they do a bit of sight-seeing; then they turn around and come back with the same procedure. The total round-trip would take about four years of Earth-time, but because of the speed at which they have been travelling, their clocks would have been running for only about two years. In other words, the astronauts will be two years younger than they would have been if they had stayed on Earth.

From their point of view, however, time on Earth will have jumped forward two years. In other words they will have travelled two years into the future.

In principle this idea is easily extendable. For example if they were to travel to a star 500 light years away and continue with the same level of acceleration, then

ABOVE: An object's movement through time and space can be represented on a "worldline". The vertical axis represents a stationary observer's movement through time. A "light cone" at 45° to the vertical represents the movement of light through space from the point of view of the stationary observer. In this diagram, space is represented by just two dimensions.

Time

Space

THE TWIN PARADOX

Astronaut Peg goes on a superfast trip round the solar system and returns to Earth five years younger than her identical twin Meg, who had been busy running experiments on the Large Hadron Collider. Peg had travelled forward in time because she had been travelling at immense speeds. But from her point of view Meg had been moving at immense speeds; so why wasn't Meg the younger of the two?

This is the twin paradox, and the answer is that during her travels, Peg the astronaut had been subject to enormous accelerations, both positive and negative, while Meg had been travelling at a relatively steady speed. The acceleration is what upsets the symmetry in their perception of time. In effect, when Peg slows at the end of her trip, stops, and starts speeding back, she jumps to a new frame of reference, so that her time no longer matches Meg's time.

they might return to Earth less than 10 years later, while time on Earth would have advanced 1,000 years. They would then have travelled 900 years into the future.

In practice we have no technology that could provide a steady acceleration of 2g for an extended period. Nor I suspect would astronauts be able to cope with that rate of acceleration indefinitely. Whether or not astronauts could survive one another's company for many years, confined in a cramped spaceship, is open to question.

In real life we are all time travellers; we all travel into the future at the rate of one second per second (1s/s). What is strange is that, in principle, we can do so at different rates. In other words time travel to the future is at least possible, thanks to Einstein's principle of special relativity. What is more, it has actually happened in practice, most obviously to space travellers.

The person who has spent the longest time in space so far is cosmonaut Sergei Krikalev, who spent 438 days in one flight on Soyuz TM-18 in the mid-1990s and a total of 803 days, or 2.2 years, in space spread over three space flights. Because his spacecraft in orbit around the Earth were travelling at about 7.6km/s (17,000mph) he has travelled about 20ms or 1/50 of a second into the future. In fact anyone who travels anywhere must travel into the future, although not very far.

BELOW: Russian cosmonaut Sergei Krikalev holds the record for time travel into the future.

ABOVE: A black hole distorts spacetime so much that nothing, not even light, can escape its gravitational pull.

OPPOSITE: A reconstruction of an event horizon surrounding a black hole. We cannot know anything about what happens beyond a black hole's event horizon as no information can cross it back towards us.

General relativity

Einstein said that the happiest idea of his life, which came to him in 1907, was that a person on Earth and an astronaut in a spaceship that was accelerating would experience the same effects. In other words gravity is the same thing as acceleration. This led him, after eight years of difficult mathematics, to his general theory of relativity, in which he explained that gravity results not from inherent mutual attraction of masses, but from the distortion of spacetime by massive objects.

Imagine a thin rubber sheet stretched across a horizontal frame. It remains horizontal until a heavy lump of metal is placed on it. Then it is distorted into a well with the lump of metal in the hollow. Roll a marble on to the sheet and it will spiral into the well, not because it is attracted by the lump of metal, but because it runs downhill on the rubber, and then (if it were a better model of the solar system) rolls round and round the hollow in constant circles. In this mini-universe the lump of metal has distorted spacetime.

Einstein's calculations showed that gravity, like travel, actually slows clocks down. The Earth's gravity makes all our clocks go slow by as much as 3ns per year, relative to (imaginary) clocks in deep space. In 1976 Robert Vessot and Martin Levine fired a hydrogen-maser clock into space, and were able to measure the fact that it gained slightly before it crashed into the Atlantic Ocean two hours later.

If the Earth were made of rubber, and could be squeezed to a fraction of its size, then gravity on the surface would increase. If you could squash it to the size of a marble (diameter 1cm or 0.4in) then clocks on the surface would stop altogether.

Cosmologist Richard Gott suggests that you could build a time machine by collecting as much mass as in the planet Jupiter, building with it an incredibly dense spherical shell just big enough not to collapse into a black hole (around 6m (20ft) in diameter), and sit in the middle of it. If anyone were to get near the outside they would be squashed flat and torn apart by the gravitational field, but inside you should feel nothing.

Any signals you send out will have to escape from the deep gravitational well caused by the shell, and the result is that outside observers will see your clock ticking at one quarter speed; so you would age only one year every four, and could therefore travel forward in time – although you might never be able to escape from your shell to enjoy the experience. You might be able to look out, however, and see the rest of the world whizzing by at four times its "normal" speed.

The ultimate time machine is a black hole, where gravity is infinite and time stops. As Paul Davies puts it, it is a one-way fast track to the end of time.

Time travel into the past

The possibility of travelling back in time creates logical paradoxes, which some people suggest make the process impossible. The best known is the "grandmother paradox": you travel back in time, meet your grandmother as a young woman, and accidentally kill her before she has given birth to your mother. Then you could

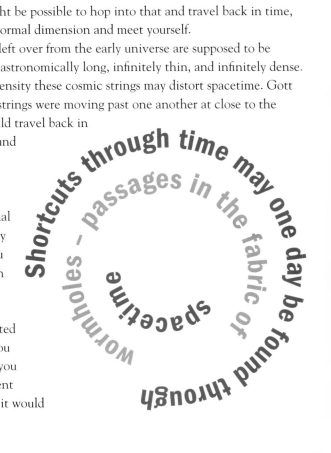

ABOVE: This graphic represents a three-dimensional cross-section through a 10-dimensional Calabi-Yau space, which string theorists theorize to exist at microscopic scales.

not be born, which means you could not travel back in time and meet your grandmother.

More simply, you might go back in time and meet a younger version of yourself; so in effect you would have cloned yourself. You should be able to remember meeting an older version of yourself from that earlier encounter, and if you were to repeat the process you might meet an even younger version of yourself – and then there would be three of you. A good plan would be to give your younger self a gold watch, for then the next time there would be two gold watches, and so on.

Also, if travel backwards in time is possible, why have we not met any travellers from the future? Surely significant events such as the assassination of President Kennedy or the birth of Jesus would have been crowded with eager spectators?

Richard Gott's answer to that question is that when a time machine is built, it may become possible to travel back to any moment after it was completed, but not before. So we have not yet met time travellers from the future because the time machine has not yet been built. When such a machine has been built, all grandmothers-to-be should beware.

One of the current attempts to create a theory of everything is superstring theory, in which elementary particles are like minuscule loops of string and our universe has eleven dimensions, rather than the three that are familiar – or four if you include time. If one of the other seven dimensions (all coiled up in tiny loops) were also a time dimension then it might be possible to hop into that and travel back in time, only to hop back into the normal dimension and meet yourself.

Some of the bits of stuff left over from the early universe are supposed to be "cosmic strings", which are astronomically long, infinitely thin, and infinitely dense. Because of their immense density these cosmic strings may distort spacetime. Gott suggests that if two cosmic strings were moving past one another at close to the speed of light, then you could travel back in time by going for a trip around the strings.

Wormholes

A slightly more conventional suggestion has been made by other physicists, that all you have to do to travel through time is to jump through a wormhole. They say that wormholes might have existed since the Big Bang, but if you jump through one of them you might re-emerge in a different universe. A safer way to do it would

Wormholes – passages in the fabric of spacetime. Shortcuts through time may one day be found through

be to make your own wormhole by causing an immense distortion of spacetime and puncturing a hole through a folded section. (Imagine the whole of spacetime as like a giant double-sided pizza. If you could tunnel through like a worm from the cheesy side to the pepperoni side the path through the tunnel would be much shorter than one that went all the way to the edge, round it, and back the other side.) Then in principle you could travel through the hole, and back by normal spaceship, and you might have gone backwards in time.

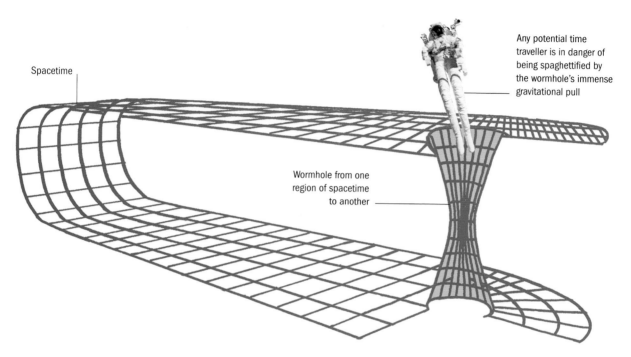

Any potential time traveller is in danger of being spaghettified by the wormhole's immense gravitational pull

Spacetime

Wormhole from one region of spacetime to another

There are problems with this plan. First, you would need more matter and energy than are available on this planet, and more technology than anyone can muster. Second, the gravitational field near a wormhole is likely to be similar to that near a black hole, and as you approached it you would almost certainly be pulled and squeezed to shreds. As your feet approached they would be pulled so hard that you would be stretched to many times your original length. At the same time your body would be squeezed to a fraction of its original thickness; in other words you would be "spaghettified". Third, you would probably be joined in the wormhole by an immense concentration of cosmic rays, which would burn you to death. I don't think I shall be queueing up to take a trip to the past.

Most of these time-travel ideas involve space and rockets and astronauts, mainly because rockets travel quickly, and one can imagine them accelerating to near the speed of light. Also because space is so big, and parts of it are so mysterious, there might well be wormholes or cosmic strings or other such weird things out there. British scientist JBS Haldane famously said: "My own suspicion is that the universe is not only queerer than we suppose; it is queerer than we can suppose."

ABOVE: In theory, it could be possible to "tunnel" through spacetime using a wormhole.

Galileo and Newton started the scientific teasing out of information about the Earth and the rest of the Universe, but it seemed that the stars were fixed in space, unmoving. Indeed until about 1800 they were called "the fixed stars". That all changed in the 1920s, when an irritating American astronomer made two great leaps forward, and showed how to calculate the age of the Universe.

THE BEGINNING OF TIME

EDWIN HUBBLE WAS IRRITATING partly because he was a wannabe Englishman. During his stay in Oxford he came to love the British lifestyle, and adopted some parts of it for himself. He took to wearing tweed jackets and smoking a pipe, and went on doing so when he went back to the US. He was said to be a difficult man to work with, but he was a great scientist.

Fellow American Henrietta Swan Leavitt had discovered that a particular type of star, the Cepheid variable, seems to have a reliably standard brightness, unlike most stars, whose brightness depends on their size. Hubble began to make a study of these Cepheid variables, especially those in the murky clouds known as spiral nebulae. No one knew what these nebulae were – clouds of gas, clouds of dust, groups of stars? – but more and more were being discovered. Hubble found that Cepheid variables in these nebulae appeared much fainter than others, and reckoned they must be fainter because they are further away.

Some stars shine with a steady brightness, but others vary: they get dimmer and then brighter again in a regular way; these are called variable stars. A well-known example is Algol (from the Arabic word meaning "ghoul").

EDWIN POWELL HUBBLE (1889–1953)

American astronomer Edwin Hubble was born in Marshfield Missouri, and at high school was an athletics star, winning seven first places in a single track-and-field meet. After completing a BSc at the University of Chicago he went to Oxford University in England as one of the first Rhodes Scholars.

He returned to the US, fought in World War I, and gained a PhD in astronomy. In 1919 he accepted a staff job at the Mount Wilson observatory near Pasadena, California, where he used the brand new 100-inch Hooker telescope, the biggest in the world at the time. When the 200-inch Hale telescope at Mount Palomar was completed he was able to use that too. Today the Hubble space telescope is a memorial to his genius.

Cepheid variables (named after Delta Cephei, a variable star observed by deaf-mute astronomer John Goodricke in 1784) are particularly interesting, both because they are extremely bright and also because their absolute brightness is related precisely to the period of variation. This means that they can be used as "standard candles". If you know how bright one really is (its absolute brightness) then you can calculate how far away it is from its apparent brightness: the dimmer it looks, the further away it must be.

ABOVE: Cepheid variables seen in the spiral arms of distant galaxy NGC 4603.

Light travels **5,878,625,373,184** miles in a year. This distance is called 1ly (light year)

At the time, most astronomers believed that there was only one galaxy – our own galaxy, the Milky Way – and that the most distant stars were a few thousand light years away. In 1924 Hubble observed Cepheid variables in the Andromeda nebula, a fuzzy patch of light visible to the naked eye, and known to astronomers for almost 2,000 years. The Cepheid variables told him that the nebula was 2.5 million light years away – much too distant to be part of the Milky Way.

Astronomical distances are so great that to quote them in kilometres or miles needs an inconvenient number of zeros. So astronomers invented big units of distance. The Astronomical Unit (AU) is the average distance of the Earth from the Sun (150 million km or 93 million miles), but even that is a bit short for objects outside our solar system. The light year (ly) is the distance light travels in one year, which turns out to be approximately 9,500,000,000,000km or 5,900,000,000,000 miles. The nearest star, Proxima Centauri, is about 4ly away, while our galaxy, the Milky Way, is about 100,000ly across.

Edwin Hubble went on to examine 22 other nebulae, and showed they were far more distant than anything in our galaxy. In other words not only did he show that these nebulae are galaxies – clusters of stars rather than clouds of dust – but at a stroke he also increased the size of the Universe by a factor of at least 10,000.

RED SHIFT

You see the stars because they emit light, and if you pass some of that light through a prism you can split it up into its constituent colours, like the colours of the rainbow. Among those colours are the spectra of hydrogen and helium, the main constituents of stars, and the lines of these spectra have known wavelengths, since they have been observed and measured in labs on Earth.

In starlight, however, these lines are moved to longer wavelengths – shifted towards the red end of the visible spectrum. This is because as the stars are rushing away from us (or in relativistic terms, as space is being stretched), each wave is strung out across a greater distance. The redshift of stars in distant galaxies shows that they are moving away from us and that the Universe is expanding. Not only that, but the Universe is expanding at an ever-increasing rate. This accelerating expansion has puzzled astronomers, who have come up with the idea of "dark energy" to account for the force that is pulling the galaxies apart.

This would have been enough of a triumph for most scientists, but Hubble went further. Learning that the light from stars is red-shifted (moved to longer wavelengths) when they are moving away from us, he went back to his galaxies, and 23 others, and looked at their red shifts. His conclusion was that all the other galaxies are moving away from us, and the more distant they are the faster they are receding; this is now known as Hubble's Law. The implication is that the whole Universe is expanding, and presumably moving away from a beginning in which it was very small.

On 28 March 1949, astrophysicist Fred Hoyle, who believed in a steady-state Universe, spoke cynically on the radio about Hubble's notion, gently mocking the notion of a "Big Bang" to begin the Universe. Using simple language to make fun of the idea to listeners, he actually coined the phrase which has now become the accepted reality: as a result of a number of measurements and inferences, we now know that it happened 13.7 billion years ago.

Scientists can run physics backwards as far as the Big Bang, but then they run into a hot dense state, a singularity, and current physics can't go no further. For the present, therefore, the Big Bang was the start of the Universe (unless one of the cosmologists' suggestions is correct, and the universe is without boundaries, with no beginning and no end). In other words Edwin Hubble's astronomical research showed us the beginning of time.

SIR FRED HOYLE (1915–2001)

A blunt and plain-speaking Yorkshireman, Fred Hoyle was born near Bradford, England, and became an astronomer at Cambridge University. He held various controversial views about cosmology, including the idea of "panspermia": that life exists throughout the Universe, and is carried about on comets or asteroids. He was the first to work out, in 1946, how heavy elements such as carbon are produced inside stars.

During the last 50 years or so there has been controversy among astronomers and cosmologists about what will happen to the Universe. There seem to be three possibilities. Either it will go on expanding forever, or it will slowly come to a stop, or it will stop and then fall inwards again, so that everything meets in a Big Crunch.

ABOVE: This is an image of the Hubble ultra deep field, showing light from galaxies billions of light years away. The Universe was only about one billion years old when some of this light was emitted.

THE END OF TIME

COMING TO A STEADY STATE SEEMS IMPROBABLE. Newton's laws say that if there are no forces acting on them then the galaxies will carry on moving in the same direction, while if there are forces then the galaxies will accelerate in the direction of the forces.

Some decades ago the general view was that the Big Bang had propelled the galaxies outwards, but the only force acting now is gravity, which should gradually overcome the outward motion, slow the movement, and reverse it, bringing the entire Universe back together in a reversal of its original motion. This should produce a Big Crunch.

Recent observations, however, suggest that the galaxies show no signs of slowing down; indeed they seem to be accelerating outwards, as though they are being pushed by some giant cosmic balloon. If this is true, then they will drift further and further apart, and space will become more and more empty.

One puzzling fact is there seems to be much more matter in the Universe than we can see. Calculations of the gravitational forces acting on some galaxies suggest that there is about 20 times as much mass in the Universe as we can account for in

The Earth will be **burned** to a crisp in 5 billion years

terms of stars. In other words, 95 per cent of the universe is made of "dark matter" (which attracts and should therefore help to bring about a Big Crunch) and "dark energy" (which repels, in a way that is not understood).

OPPOSITE: In five billion years' time, the Sun will swell up into a red giant, expanding to engulf the Earth, at which point life on Earth as we know it will come to an end.

As for the inhabitants of Earth, the ultimate fate of the Universe is somewhat academic, because our time will end relatively soon. The Sun will gradually cool and expand to hundreds of times its present size, becoming a red giant. In the process it will swallow up Mercury, Venus, and probably also Earth; even if Earth is not swallowed up, the planet will still be burned to a crisp. We don't need to panic immediately, however, since this is not going to happen for about five billion years.

A nanosecond (ns) is one billionth of a second, but this is still a long time compared with the femtosecond, the attosecond, and the shortest possible unit of time, Planck time.

THE SHORTEST SEGMENTS OF TIME

IN BETWEEN A NANOSECOND AND PLANCK TIME comes the femtosecond (fs), which is one millionth of a nanosecond. The peaks in a light wave arrive at your eye a few femtoseconds apart. The shortest unit of time yet measured is the attosecond (as). In May 2010 German scientists measured a time interval of 12as.

Superfast stingers

Jellyfish can sting fast, but no one knew quite how fast, until a team of scientists at the University of Heidelberg managed to film the process with an amazing "streak" camera, taking pictures at the astonishing

MIDDLE: A jellyfish's sting has an acceleration of up to 5 million g, giving it the same sort of penetrating power as a bullet.

There are more *attoseconds* in one second than there have been seconds in the entire life of the Universe

IMAGINING A NANOSECOND

The speed of light – the "speed limit of the universe" – is the speed at which light travels through a vacuum. In one nanosecond, light travels exactly 29.97992458cm, or just under a foot.

In 1ns, light travels about a foot, 30cm, or this far...

| 0 | 5 | 10 |

PLANCK TIME

Planck time, 10^{-43}s, is the basic unit of time in the theory of quantum gravity; it is defined as the time light takes to travel the Planck length, 4×10^{-35}m. This is the smallest unit of time that makes any sense in normal physics. In effect, the Universe began not at time zero, but one Planck time later. At the quantum scale, the position of particles such as electrons is uncertain, and this uncertainty is described by the particle's "wave function". For the equations of gravity to work, there needs to be a gap between the wave functions of different particles (top). If the peaks of the wave functions are less than a Planck length apart, the meaning of gravity as currently understood breaks down; we can have no information about what is going on.

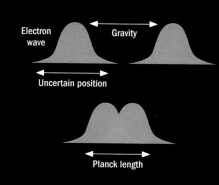

rate of around 1.4 million per second. When the jellyfish detects prey it releases a poison-filled cell called a nematocyst within 700ns, and then fires a dart (a "stylet") to puncture the skin of its victim and inject the poison. The jellyfish's sting is one of the fastest movements in the animal kingdom; it is catapulted with an acceleration up to 5 million g.

The birth of planets

The secrets of the giant planets may soon be explored by scientists at the National Ignition Facility in California, who plan to fire a colossal burst of energy into a tiny ball of iron. Their ultraviolet lasers can deliver 500 million megawatts of energy in 20 ns, which inside the millimetre-sized ball will vaporize the iron, creating a colossal shock-wave that should reproduce some of the processes involved in the formation of the cores of such planets as Jupiter and Saturn. The same technology may later be used to induce nuclear fusion of hydrogen, a process that could solve the world's energy problems.

SMALLEST TIME SEGMENTS

Planck time	10^{-43} seconds
attosecond (as)	10^{-18} seconds
femtosecond (fs)	10^{-15} seconds
nanosecond (ns)	10^{-9} seconds
microsecond (μs)	10^{-6} seconds
millisecond (ms)	10^{-3} seconds

ABOVE: An attosecond is one quintillionth of a second, or a billionth of a billionth of a second, or a millionth of a millionth of a millionth of a second.

... that's about 300,000 kilometres per second, or 186,000 miles per second

INDEX

ACKNOWLEDGEMENTS

Picture credits:

8-9 Ton Koene/ZUMA Press/Corbis, 10-11b The Gallery Collection/Corbis, 10tl DEA Picture Library/Getty, 12tl Andy Crawford/Getty, 12c BananaStock/Thinkstock, 13 Sean Gladwell/Dreamstime.com, 14-15 Steve Allen/Dreamstime.com, 15br Mike.lifeguard/Wikicommons, 16bl 2011 Thinkstock, 16-17c Blackslide/Dreamstime.com, 17tr Theo Gottwald/Dreamstime.com, 18bl Tenzintrepp/Wikicommons, 19 Nacivet/Getty, 20tl Solarseven/Dreamstime.com, 20-21c Alex Borisovich/Dreamstime.com, 21tr Jody Wissing/Dreamstime.com, 22 Torian Dixon/Dreamstime.com, 23c Buddy Mays/Alamy, 24tl Wade Iedema/Dreamstime.com, 25r Adeliepenguin/Dreamstime.com, 26-27 Ryan Carter/Dreamstime.com, 26bl Cammeraydave/Dreamstime.com, 28bl David Ball/Corbis, 29tr Bhairav/Dreamstime.com, 29br Galagorn/Wikicommons, 30bl Blavarg, Susanna/Getty, 31 William Attard McCarthy/Dreamstime.com, 32tr 2011 Thinkstock, 33tl Blotty/Dreamstime.com, 33tl Damiens.rf/Wikicommons, 34bl Norberto Mario Lauría/Dreamstime.com, 35tr 2011 Thinkstock, 36-37 Gregor Kervina/Dreamstime.com, 38cl Sakala/Shutterstock , 39tr Martin Green/Dreamstime.com, 39br 2011 Thinkstock, 40tl Budda/Dreamstime.com, 40bl Cammeraydave/Dreamstime.com, 41br 2011 Thinkstock, 42b Ton Koene/ZUMA Press/Corbis, 43tr 2011 Thinkstock, 44-45 Monkey Business Images/Dreamstime.com, 44tl Trekphiler/Wikicommons, 46tl TTaylor/Wikicommons, 46-47b Thomas Pozzo Di Borgo/Dreamstime.com, 48bl El Comandante/Wikicommons, 49tr Rinusbaak/Dreamstime.com, 49br Juliux/Wikicommons 50bl NASA, 50-51c NASA, 51tr Alberto Fernandez Fernandez/Wikicommons, 51br NASA/WMAP Science Team, 52-53t Kathy Collins/Getty, 54tl Vinicius Tupinamba/Dreamstime.com, 54-55b Kennan Ward/Corbis, 55br Tau_olunga/Wikicommons, 56tr Oriontrail/ Dreamstime.com, 56bl NASA, 56br NASA, 57br www.michaelmarten.com,58tr John Foxx/Getty, 59tr Getty Images, 59br Luc Viatour/Wikicommons, 60tl 2011 Thinkstock, 60br Comstock Images/Thinkstock, 61c 2011 Thinkstock, 62l Adam Hart-Davis, 63tr Jupiterimages/Thinkstock, 64tl Dcoetzee/Wikicommons, 65tr Adam Hart-Davis, 64-65b Goce Risteski/Dreamstime.com, 66-67 Roy Toft/Getty, 66bl SSPL/Science Museum/Getty, 67tr History of Science Collections, University of Oklahoma Libraries, 68tl Dcoetzee/Wikicommons, 69tr Laurascudder/Wikicommons, 69br Lillian Obucina/Dreamstime.com, 70bl Diego Vito Cervo/Dreamstime.com, 71tr Digital Vision/Thinkstock, 71br TomCatX/Wikicommons, 72-73 Arvind Balaraman/Dreamstime.com, 74br SSPL via Getty Images, 75tr Tom Brakefield/Thinkstock, 75br Quibik/Wikicommons, 76tl Seav/Wikicommons, 76bl Ephraim33/Wikicommons, 77b Antoine Beyeler/Dreamstime.com, 78-79 Daryl Balfour/Getty, 78tl Archaeodontosaurus/Wikicommons, 80c Mauswiesel/Wikicommons, 81b 2011 Thinkstock, 82bl 2011 Thinkstock, 83 Richard A. Cooke/Corbis, 84b Roger Ressmeyer/Corbis, 85tr Jesús Eloy, Ramos Lara/Dreamstime.com, 85b HereToHelp/Wikicommons, 86tl 2011 Thinkstock, 86-87c Markus Gann/Dreamstime.com, 87tr Jacks Rache/Wikicommons, 88-89b Edurivero/Dreamstime.com, 88tl Hemera Technologies/Getty Images, 89tr 2011 Thinkstock, 90bl Tatyana Chernyak/Dreamstime.com, 91cr Wavebreakmedia Ltd/Dreamstime.com, 91cl Wavebreakmedia Ltd/Dreamstime.com, 92-93 Andreas Weber/Dreamstime.com, 92br Cammeraydave/Dreamstime.com, 93bl Richard Baker/In Pictures/Corbis, 94t Tomas Hajek/Dreamstime.com, 95b Natalie Fobes/Getty, 96-97b Stephen Alvarez/Getty, 96tl Bettmann/Corbis, 98-99 Bricktop NASA, 99tr NASA, 100 George Shuklin/Wikicommons, 100-101 Vladimír Ondr__ek/Dreamstime.com, 101br Patrick Durand/Sygma/Corbis, 102b Daniel J Cox/ Getty, 103bl NASA, 103tr Ying Feng Johansson/Dreamstime.com, 104tl Njari/Dreamstime.com, 104bl Zimbres/Wikicommons, 105 Alexandr Ozerov/Dreamstime.com, 106br ArtMechanic, 107t 2011 Thinkstock, 108-109b Kateleigh/ Dreamstime.com,108tl 2011 Thinkstock, 109tr Philip J Brittan/Getty, 110bl NASA, 110b NASA, 111 Jamen Percy/Dreamstime.com, 112-113t Hulton-Deutsch Collection/Corbis, 114l 2011 Thinkstock, 114-115 DAVID MERCADO/Reuters/Corbis, 116-117b IIC/ Axiom/ Getty, 116tl Euchiasmus/Wikicommons, 118tl 2011 Thinkstock, 119 Skyscan/Corbis, 120-121 Dexter Lane/ Getty, 122c Shariff Che' Lah/Dreamstime.com, 123tr Tony Hallas/Getty, 123br 2011 Thinkstock, 124br Leinad-Z/Wikicommons,126tl Albert1ls/Wikicommons, 126t Albert1ls/Wikicommons, 126-127 Saniphoto/Dreamstime.com, 127c FunkMonk/Wikicommons, 128tr 2011 Thinkstock, 128-129 Roger De La Harpe; Gallo Images/Corbis, 130-131 Lee Frost/Getty, 131tr 2011 Thinkstock, 132b Ammit/Dreamstime.com, 133tr Nadezhda Bolotina/Dreamstime.com, 133br DEA/A. DAGLI ORTI/Getty, 135 J. Paul Getty Museum, Los Angeles, USA/The Bridgeman Art Library, 136tl Mattana/Wikicommons, 136br 2011 Thinkstock, 137tr SSPL via Getty Images, 138t Eloquence/Wikicommons, 139tr Ammar Awad/Reuters/Corbis, 139b 2011 Thinkstock, 140tl Thomas Lozinski/Dreamstime.com, 140bl Magnus Manske/Wikicommons, 141b Kazuyoshi Nomachi/Corbis, 142-143 Andesign101/Dreamstime.com, 143t Andrew Dunn, 144-145 Werner Forman/Corbis, 145tr Madman2001/Wikicommons, 145br Rosemania/ Wikicommons, 146 Dabobabo/Dreamstime.com, 147tr Bcody80/Wikicommons, 147b Paul Hakimata/ Dreamstime.com, 148 Rama/Wikicommons, 149t Joe Kress/Wikicommons, 149r Gaja/Dreamstime.com, 150b SSPL via Getty Images, 151tr Thomas Jenkins/Dreamstime.com, 151br Getty Images, 152l CHIP EAST/Reuters/Corbis, 153 Baloncici/Dreamstime.com, 154-155 SSPL/National Railway Museum, 154tr CBX/Wikicommons, 156-157 2011 Thinkstock, 156b 2011 Thinkstock, 157br Public Domain, 158b SSPL via Getty Images, 158bl Mahahahaneapneap/Wikicommons, 159t Bettmann/Corbis, 160tr Dcoetzee/Wikicommons, 160bl Eubulides/Wikicommons, 161b 2011 Thinkstock, 162-163 Vicki France/Dreamstime.com 164cl Jamen Percy/ Dreamstime.com, 164-165c NASA, 165tr NASA, 165tr NASA 166-167 2011 Thinkstock, 168br Pascalou95/Dreamstime.com, 169r Sebastian Le Clerc/Wikicommons, 170tl Adam Hart-Davis, 170br Adam Hart-Davis, 171r Al-Jazari/Wikicommons, 172tr Roman/Getty,172bl French School/Getty, 173t S711/Wikicommons, 174cl Heliocrono/Wikicommons, 175br Nigelj/Wikicommons, 174-175 David Abbott/Dreamstime.com, 176bl Cosmin - Constantin Sava/Dreamstime.com, 177bl Chetvorno/Wikicommons, 177tr Leinad-Z/Wikicommons, 178 Tim Martin/Dreamstime.com, 179tr Nikos Pavlakis/Dreamstime.com, 179bl VASSILIS PSOMAS/epa/Corbis, 180br SSPL via Getty Images, 181r Wessel Cirkel/Dreamstime.com, 182-183 Wieslaw Jarek/Dreamstime.com, 182br Dmitry Rozhkov/Wikicommons,183tl Kg Kua/Dreamstime.com, 183bl SSPL/Science Museum, 184l SSPL via Getty Images, 185tr Chetvorno/Wikicommons, 186bl Timwether/Wikicommons, 186t Zanyjazz/Wikicommons, 187br SSPL via Getty Images, 188-189 Kacpura/ Dreamstime.com, 188l SSPL via Getty Images, 189tr Sergei Gutnikov/Wikicommons, 189bl Wrs1864/Wikicommons, 190tl SSPL/Science Museum, 190br Elimitchell/Dreamstime.com, 191tr Myotus/Wikicommons, 192bl SSPL/Science Museum, 193tr SSPL/Science Museum, 195b National Maritime Museum, 196tl Dmitry Rozhkov/Wikicommons, 196br Underwood & Underwood/Corbis, 197tr National Maritime Museum, 198l SSPL via Getty Images, 199tr SSPL/Science Museum, 199b SSPL/Science Museum, 200tl Chribbe76/Wikicommons, 200bl Archaeodontosaurus/Wikicommons, 201tr SSPL via Getty Images, 201br Francis Flinch/Wikicommons, 202-203 Pseudolongino/Dreamstime.com, 202br STFC, 203bl NASA; vectors by Mysid, 204tl Petr Vaclavek/Dreamstime.com, 204br Dnn87/Wikicommons, 205tr Svdmolen/Wikicommons, 205b James Stevenson/Getty, 206b Dennis Thompson/Dreamstime.com, 207br Hannu Viitanen/Dreamstime.com, 207tr Dieter Spannknebel/Getty 208b Pniesen/Dreamstime.com, 209tr Kts/Dreamstime.com, 209br Antonuk/Dreamstime.com 210-211 NASA, 212bl EPO/Wikicommons, 213c NASA, 213br Dcoetzee/Wikicommons, 214tr Gianni Dagli Orti/Corbis, 214bl Aushulz/Wikicommons, 215 Fir0002/Wikicommons, 216-217 Stephen Mcsweeny/Dreamstime.com, 216bl Juriah Mosin/Dreamstime.com, 218c Digital Vision/Thinkstock, Freedomimage/Dreamstime.com, 218-219 Davidstockphoto/Dreamstime.com, 220bl Glj1952/Wikicommons, 221 Tall Tree, 222-223 Geopappas/Dreamstime.com, 222 tl Vladimir Il'yin/Dreamstime.com, 223b Alejandro Duran/Dreamstime.com, 224b Martin Novak/Shutterstock, 225r 2011 Thinkstock, 226-227 Evgeny Prokofyev/Dreamstime.com, 228 Adam Hart-Davis, 229t Robert Young/Dreamstime.com, 229b Antilived/Wikicommons, 230bl Petr Malyshev/Dreamstime.com, 231-232 NASA, 232bl Hemulen/Wikicommons, Quibik/Wikicommons, 234br Graham Bloomfield/Wikicommons, 236b Martial Trezzini/epa/Corbis, 237tr Alexandr Mitiuc/Dreamstime.com, 239tr 2011 Thinkstock, 239br NASA, 240tl D'ARCO EDITORI/Getty, 241 CorvinZahn/Wikicommons, 242tl Lunch/Wikicommons, 243tr Bricktop/Wikicommons, 244br University of Chicago, 245 NASA, 246-247 Jmencisom/Wikicommons, 247br St John's College Library, 248tl JJ Harrison/WIkicommons, 249 Mark Fisher/Aurora Photos/Corbis 250-251c Kelpfish/Dreamstime.com

Thanks to:

I would like to thank my wife Dr Sue Blackmore for her patience and for innumerable helpful suggestions, Peter Taylor and David Lamb who provided much early encouragement and advice, and Professor Sir Michael Berry, Professor Ernst Zürcher, and Deborah Hutchinson for reading and improving various parts of the manuscript. I have no doubt managed to include a number of errors in spite of their help. Thanks also to Tonewoood (Switzerland) for their advice, and to Jonathan Betts at Greenwich and David Thompson at the British Museum for their time and kindness in talking to me. I would also like to thank David John, Rob Colson, and the rest of the team at Tall Tree Ltd for their splendid work in sourcing the illustrations and designing the book.